Russia's Coercive Diplomacy

Russia's Coercive Diplomacy
Energy, Cyber, and Maritime Policy as New Sources of Power

Ryan C. Maness
Visiting Fellow of Security and Resilience Studies, Northeastern University, Boston, MA

Brandon Valeriano
Senior Lecturer, University of Glasgow, UK

© Brandon Valeriano and Ryan C. Maness 2015.

All rights reserved. No reproduction, copy or transmission of this publication may be made without written permission.

No portion of this publication may be reproduced, copied or transmitted save with written permission or in accordance with the provisions of the Copyright, Designs and Patents Act 1988, or under the terms of any licence permitting limited copying issued by the Copyright Licensing Agency, Saffron House, 6-10 Kirby Street, London EC1N 8TS.

Any person who does any unauthorized act in relation to this publication may be liable to criminal prosecution and civil claims for damages.

The author(s) has/have asserted his/her/their right(s) to be identified as the author(s) of this work in accordance with the Copyright, Designs and Patents Act 1988.

First published 2015 by
PALGRAVE MACMILLAN

Palgrave Macmillan in the UK is an imprint of Macmillan Publishers Limited, registered in England, company number 785998, of Houndmills, Basingstoke, Hampshire, RG21 6XS.

Palgrave Macmillan in the US is a division of St Martin's Press LLC, 175 Fifth Avenue, New York, NY 10010.

Palgrave is the global academic imprint of the above companies and has companies and representatives throughout the world.

Palgrave® and Macmillan® are registered trademarks in the United States, the United Kingdom, Europe and other countries.

ISBN: 978–1–137–47943–3

This book is printed on paper suitable for recycling and made from fully managed and sustained forest sources. Logging, pulping and manufacturing processes are expected to conform to the environmental regulations of the country of origin.

A catalogue record for this book is available from the British Library.

A catalog record for this book is available from the Library of Congress.

*Dedicated to Ted and Karen Maness and
to those who have devoted their lives to the analysis of Russia*

Contents

List of Figures and Tables viii

Preface x

Acknowledgements xiii

1 Introduction: New Forms of Coercive Power in the Putin Era 1

2 Russia's Foreign Policy Choices and the Application of Situational Coercive Diplomacy 21

3 Rivalry Persistence and the Case of the United States and Russia: From Global Rivalry to Regional Conflict 45

4 Russia in Cyberspace 85

5 Russian Coercive Energy Diplomacy in the Former Soviet Union 108

6 Energy Salience and Situational Coercive Diplomacy: Comparison of Coercive Energy Policy in the Caucasus and Central Asia 138

7 Russian Foreign Policy in the Arctic: Regional Issues, Preferences, and Conflict Moderation 173

8 Conclusion: Russian Coercive Diplomacy after the Cold War 205

Notes 213

Index 243

List of Figures and Tables

Figures

1.1	Russian and US GDP growth comparison 1997–2008	10
1.2	Russian exports and imports 1994–2007	12
1.3	Yearly percentage change in Russian military expenditures	14
2.1	Sources of foreign policy outlook	36
3.1	Russian opinion of the United States over time	67
3.2	American overall views of Russia	71

Tables

3.1	Russian opinion of US treatment of Russia	68
3.2	Russian opinion of states it considers unfriendly or hostile	69
3.3	American opinion of Russia	70
3.4	American opinion on Russian foreign policy	71
3.5	Russian opinion on the motives and purpose of NATO	73
3.6	Russian opinion on whether NATO members in Europe should fear Russia	73
3.7	Russian opinion of Ukraine and Georgia joining NATO	74
3.8	Russian opinion about Russian military intervention in Georgia in 2008	75
3.9	Russian opinion of threatened NATO airstrikes on Bosnian Serbs	76
3.10	Russian opinion of NATO intentions in the Kosovo crisis of 1999	77
3.11	Russian opinion of NATO bombings of Serbia during the Kosovo crisis	78
3.12	Russian opinion of the NATO intervention in Libya	79
3.13	Russian opinion of the cause of the Syrian conflict	81
4.1	Overall cyber capabilities among states	93
4.2	Summary of cyber incidents involving Russia 2001–11	96
4.3	Number of Russian Cyber Interactions compared to Other Cyber Powers	104
5.1	Comparative natural gas prices of the former soviet union in 2005	119
5.2	Russian economic statecraft with natural gas prices ($ per mcm)	129

List of Figures and Tables ix

5.3	Russian economic statecraft with natural gas prices ($ per mcm) with dyadic dummies	133
6.1	Natural gas salience scores between Russia and former Soviet States (Low 0–4, Moderate 5–9, High 10–13)	146
6.2	Former USSR regional sub-complex salience scores	146
6.3	Top five FDI energy investors in Azerbaijan ($/million)	150
6.4	Russian opinion of states it considers unfriendly or hostile in the caucasus	151
6.5	Russian opinion of the cause of the August 2008 Russian-Georgian conflict	152
6.6	Russian opinion of Western support of Georgia in the August 2008 conflict	152
6.7	Natural gas pricing in the Caucasus Region ($ per mcm)	155
6.8	Existing Caucasus natural gas pipeline	157
6.9	Proposed Caucasus natural gas pipelines	158
6.10	Chinese natural gas agreements with Central Asian states	161
6.11	Russian opinion of reliable international partners in Central Asia	164
6.12	Russian opinion of Central Asian states that are the most politically stable and economically successful	164
6.13	Russian opinion of Russian-Chinese relations	164
6.14	Russian opinion about the future of Russian-Chinese relations	165
6.15	Natural gas pricing in the Central Asian region ($ per mcm)	168
6.16	Existing Central Asian natural gas pipelines	169
6.17	Proposed Central Asian natural gas pipelines	170
7.1	Dyadic salience scores for Arctic issues (Low 0–7, Moderate 8–13, High 14–20)	185
7.2	Arctic regional factors of the Arctic Five	192
7.3	Preferred approach to resolving Arctic disputes	198
7.4	Awareness of Arctic Council	201
7.5	Support for Arctic Council	201
7.6	Support for expanded Arctic Council mandate	202

Preface

This book on Russian foreign policy engagements after the Cold War is in many ways a successor to my book on the origins of rivalry (*Becoming Rivals: The Process of Interstate Rivalry Development*, 2013). In *Becoming Rivals*, I explain how rivalry develops. This typically occurs through the use of power politics foreign policy strategies. An unanswered question is why states use coercive diplomacy in the first place. Once these strategies start a rivalry, they also become learned behaviors and thus perpetuate a rivalry. What remains is to ask, What would lead a state to choose the more belligerent path in the foreign policy realm? Where do power politics strategies come from, and why are they perpetuated? What happens when a state is engaged in a rivalry and continues to utilize power politics practices?

Some theories give too simple of an answer to these questions. They might suggest that coercive diplomacy is used because it supports the national interest or because it is expected of a powerful state. Not asserting a state's power makes it seem weak and cowardly. Yet this process does not explain how the power politics situation develops and what choices states make. We argue here that coercive diplomacy is used by Russia when there is an ongoing rivalry in the region of focus, when public opinion supports such options, and when the issue has a high level of salience. To examine salience further, one must break down the issue into its component parts, but also understand how timing, regions, and the public and elites feel about a particular issue.

To understand international diplomacy, one must peel back the levels to examine action. There are contexts that determine choice and, further, the contexts such as an issue at stake have their own particular dimensions that must be examined. We undertake this complex task in this volume. The goal is to explain Russian foreign policy choices. We argue that Russia uses coercive diplomacy in post-Soviet space because of the past history of disputes in the region, the high salience of issues at stake, issues of Russian great power identity, and because the public supports these actions. Russia also challenges the United States in this region based on these characteristics. Power politics strategies can and will be used by states, but to understand how these choices are made one must not look to the structure of the system or compulsion but to the characteristics of the dispute under examination and the values of the actors involved.

A great intellectual debt in this volume is owed to John Vasquez since his work has guided the work in this volume and also in our careers. He is known for three main contributions to international relations: the issue-based paradigm, the steps-to-war research program, and using various approaches to criticize the impractically and lack of evidence behind the realist paradigm. Each of these contributions directly leads to this volume. In many ways, this work is a criticism of realism as a universal principle or a guide to foreign policy. Strategies realists often support are used in foreign policy, but not for the reasons theorists suggest. Context matters; the situation and values of states determine action and choices rather than pure power politics calculations. What is more, the choices realists suggest in the realm of foreign policy are often not evaluated for effectiveness. We argue here that when Russia uses coercive diplomacy, it often fails to achieve its objectives.

This book also follows the issue-based approach in focusing on the issues central to state-to-state relations. Rather than examining the entire course of relations, scholars should focus on the issues under contention to examine how foreign policy tools are deployed and developed. Some issues are more contentious than others, mainly territorial ones. Other issues motivate the public to push for action. The issue at stake should be the primary focus of foreign policy scholars because issues motivate action. This book is a call for a return to this foreign policy perspective. Other perspectives are valuable and add to the complete picture of what makes a foreign policy portfolio, but the issue at state should be included in any analysis of the motivations and outcomes of foreign policy processes.

Russia, in many ways, was chosen as the unit of interest in this book because of the plethora of foreign policy concerns and situations it involved itself in after the Cold War. While the United States is now the clear global hegemon, it also has some responsibility to act in a restrained manner so as to not provoke a coordinated action against its interests. Russia has no such check on its use of power; it can and will use force wherever it is able in order to protect its historic interests. More importantly, Russia represents a transitioning state in terms of its political system, but also its foreign policy system after the Cold War. Its interests are vast, but mainly located in the post-Soviet region. The breath of this study comes not from the variation in our cases, but the variation within the case. Russia has activities in just about every realm of foreign policy. Its rivalries and issue conflicts are many, leaving the region a ripe area to study. Russia also represents an important state to examine for its geopolitical value. Most examinations of Russia foreign policy focus on specific contexts or cases rather than the totality of Russian interests. Others also fail to examine the outcomes of foreign policy choices, instead focusing on the tactics used. Here we examine both, but we also examine these issues in new contexts, such as energy, cyber, and maritime issues in the Arctic.

We dedicate this book in part to Russia scholars such as Donna Bahry, Stephen White, and Walter Zimmerman, whose long interest in the region has informed our analysis. This work is a combination of a regional study with specific international relations theories and practices that have become the norm in the field. It is informed by and uses quantitative analysis to get at the question of just what Russia is doing in the foreign policy realm, why, and to what end. Without the knowledge and advances of Russian scholars, this work would never have gotten off the ground.

In many ways, we are interlopers into the region. Russia as a case serves our interests rather than guiding our interests. In my view, this is how scholarship should be done in the international relations field. We choose our case based on the questions we wanted to ask and the tools available rather than our case choosing the questions we ask. Many of the theories utilized in this book were developed through quantitative investigations. The combination of qualitative and quantitative theory, on one hand, and regional analysis, on the other, should not be ignored. Empirical theories should be applied and evaluated according to regional units of interests. But we are agnostic towards the focus on one region or another. Instead we follow the paradigms that interests us, specifically the issue-based approach, constructivism with a quantitative focus, and the steps-to-war approach in examining conflict.

We hope you find this volume interesting and useful. We also hope that this volume can be a guide to the kinds of studies needed in the international relations field. Theory and analysis is useless unless one is able to apply it to a specific context or current reality. Motivations and outcomes should be examined, rather than left to others to develop. This book is built on the backs of others whose insights have guided us to this point.

Brandon Valeriano
Glasgow, UK
August 2014

Acknowledgments

This book is written because of a lifetime of interest in Russia—an interest that goes back to my childhood living in Europe in the early 1980s. For five years, I lived in Belgium near the frontlines of the Iron Curtain that divided east and west for so many years. With the peaceful end to the Cold War, my interest in Russia and its relations with the rest of the world never waned. After majoring in political science and history at the University of Illinois at Urbana-Champaign, my entry into the graduate program in the Department of Political Science at the University of Illinois in Chicago has brought me to the point where a lifetime of interest and research in Russian foreign policy has culminated into the pages of this book.

This book is an outcropping of my first publication, "Russia and the Near Abroad: Applying a Risk Barometer for War" (with Brandon Valeriano) in the *Journal of Slavic Military Studies*, 25 (2): 125–148, as well as my doctoral dissertation "Coercive Energy Policy: Russia and the Near Abroad," defended at the University of Illinois at Chicago. We expanded the knowledge gained from these projects into this volume.

Many thanks go out to three professors at UIC: Andrew McFarland, Yoram Haftel, and Dennis Judd. Professor McFarland's constant support of my work and interests kept me progressing towards my doctorate. Professor Haftel introduced me to the subfield of international political economy, a field that I have applied to my research in many projects. Finally, Professor Judd always encouraged and complimented my writing style, and his teachings in how to get published by writing well have made this book possible.

I would also like to thank my parents, Ted and Karen Maness, who always encouraged me to follow my interests and to do what makes me happy. Their outstanding support and parenting have given me the drive and perseverance that have allowed for my success. Now in a better place and no longer on this Earth, I know that they are looking down on me with pride.

I therefore dedicate this book to my family, whom I love more than anything else.

Ryan C. Maness
Chicago, IL
August 2014

1
Introduction: New Forms of Coercive Power in the Putin Era

The Setting

Russian power and its projection has been a key concern for international interactions since the birth of the modern state system. The downing of Malaysia Airlines flight 17 over Eastern Ukrainian territory on 17 July 2014 brought the ire of the international community upon President Vladimir Putin's Russian government.[1] British Prime Minister David Cameron likened Russia's actions, which include supplying the ethnic Russian separatists fighting the Ukrainian government as well as annexing the Crimean peninsula from Ukraine in March 2014, with the early warning signs that sparked both world wars of the 20th century. "In a way, this is what we see today in Europe. Ukraine is a country recognized by the United Nations, a country which has and should have every right to determine its own future. . .it has the right not to have its territorial integrity impugned by Russia."[2] United States President Barack Obama, referring to the separatists whom he acknowledged were being supported by Russia, called the tragedy "an outrage of unspeakable proportions."[3]

The commercial airliner was flying over a part of Ukraine controlled by ethnic Russian separatists, and all 298 passengers were killed. According to Western intelligence, the airplane was shot down by Russian surface-to-air missiles, supplied by Russia, from territory controlled by these rebels.[4] Although Russia has vehemently denied its part in this tragedy, the situation in the Crimea is clear.[5] The ouster of pro-Russian President Viktor Yanukovych in February 2014, which nullified the planned economic partnership between Ukraine and Russia, led to Russia's annexation of Crimea from Ukraine in March. The country has since deteriorated into a civil conflict where Russia has been accused of arming and funding the separatist faction in the breakaway region of

Donetsk. Russian involvement in the destabilization of Ukraine and the evidence supporting its part in backing the ethnic Russian separatists has become apparent, and economic sanctions from the United States and European Union (EU) have been put in place.[6]

Why would Russia behave in such a coercive manner with a much weaker neighbor? Why would it annex sovereign territory of another country, and why would it want the Ukrainian government troubled with armed and dangerous rebel factions challenging its rule? Why would it cut off gas supplies to the country in the midst of winter? What makes Ukraine particularly salient to Russia such that it would risk rising tension and blockade from Western governments? The purpose of this volume is to uncover the development and process of Russia's recent coercive behavior through the prism of what we call new forms of power—primarily cyber, energy, and legal/institutional maneuvers.

International interactions are undoubtedly changing, but often the context and content of this change has not been analyzed from what might be called the modern application of the power politics perspective. Some define politics as the authoritative allocation of power.[7] Under this framework, our study here examines power politics as tactics shift and develop in light of new methods. We focus on how these new forms of power are used with a focus on Russia's application in the post-Cold War system. We are not suggesting that power has transformed, that soft power dominates—our story is that the contours of international interactions have shifted, but the outcome remains the same.[8] New forms of power as seen by cyber tactics, the use of energy politics in the realm of gas and other hydrocarbons, and legal and naval maneuvering on the seas are being used by traditional powers in new ways. We evaluate the efficacy of these forms of power but also seek to change things up and deviate from traditions in specific fields. This work challenges the assumptions many have about the forms of power, international interactions, and the outcomes of these maneuvers.

We seek to analyze the consequences of a Russia struggling with its identity as a midrange power after the end of the Cold War. We fill an important gap in the literature by exploring the reemergence of Russia through the use of new power politics tactics in the realm of cyber security, energy security, and maritime power. This is not a book about Russia so much as it is a book about how a revitalized power such as Russia uses new opportunities in light of constraints. Russia cannot compete on the traditional material power battlefield, even in Ukraine it has used covert tactics rather than overt power, and thus its tactics have shifted. How do these shifting Russian foreign policy practices impact the course of peace and stability in the Eastern European, Caucasus, and Central Asian regions of post-Soviet space, as well as the Arctic region? Will cyber conflict dominate in the region and be a tool of the Russian foreign policy regime? How does Russia use its large endowments of hydrocarbons to its advantage? What is Russia's maritime Arctic policy as the ice caps continue to melt? These are all important questions that this volume will tackle.

We delineate the implications of Russian foreign policy actions on its neighbors and the United States through the use of developing power techniques. What does a reassertive Russia tell us about the importance of issues of disagreement, the new sources of power, and conflict at stake between states? We hope to tell the story of what Russia's goals are, how it goes about achieving its goals, and the consequences of foreign policy action through a theory we call "situational coercive diplomacy." This belligerent and assertive foreign policy decision making is dangerous; however, Russia uses power politics tactics only when salient issues are at stake, the public supports such actions, and there is a historical rivalry process at work.

The most important goal of this research is to understand how developing sources of power are utilized in the modern context. Russia's place in the world as a powerful regional actor can no longer be denied; the question left is, what does this mean in terms of foreign policy and domestic stability for the actors involved in the situation as Russia comes to grips with its newfound sources of might? Russia is a useful case since it seems to be the standing example of how modern forms of power blend with traditional coercive tactics. Our question is what the outcome of these tactics is; do the means achieve their desired ends?

The Russian Context

As Russia reemerges from a long slumber after the demise of the Soviet system, it has begun to engage the world through the use of power politics tactics and coercive diplomacy once again. This engagement is of a different sort than its use of power during the Cold War. Much to its dismay, Russia is no longer a global superpower. It has broad interests and acts across the globe just like any other major power, but its power projection capabilities are limited mostly to the region made up of the former republics of the Soviet Union.[9] Russia can no longer hope to be the global power that challenges the United States, but it can influence American actions in its post-Soviet backyard and a bit beyond through the use of power politics tactics. We also note that at times Russia uses accommodation to achieve cooperation on issues that are not as salient as those that deal with Russian ethnic minorities or specific territorial claims.

Examples of Russian increased political, economic, and military involvement in post-Soviet space are replete: support for Armenia in the disputed Nagorno-Karabakh region of Azerbaijan, support for the separatist enclaves of Abkhazia and South Ossetia in Georgia and Transdniester in Moldova, and the Russian military bases in these troubled spots in the Russian sphere of influence. Armenia has found itself close to Russia since the Soviet breakup. Azerbaijan, a predominately Muslim country, has not been able to regain its sovereignty over the disputed region due to the continuing supply of arms by Russia to Armenia. Russian military bases are also found in the breakaway regions of Georgia and Moldova; and these states have also not been able to govern their own territories due to the presence of the more powerful Russian military within their borders. The importance of this analysis is reinforced even more by recent events in Ukraine, where the government in Kiev has rejected a closer relationship with the European Union in favor of an economic boost from Russia. Moscow has offered huge subsidies in natural gas pricing that seems to have been too good of a deal for Ukraine to move closer to the West, and it is now inching toward the Russian political orbit.[10]

The issuance of Russian passports to ethnic Russians in post-Soviet states, the subsidization of hydrocarbons for Moscow-friendly states, the assets-for-debts programs in order to collect debts from these states, and the supposed tampering in the elections of some of the states of the region are also other foreign policy tactics used by the Russian state to exert its power in its former empire. Often these policies are met with diplomatic opposition from the West and specifically, the United States. Some Near Abroad states (Belarus, Armenia, Kyrgyzstan, and Tajikistan) have embraced this increased Russian involvement, others have abhorred it (Georgia, Estonia, Lithuania, and Latvia), and still others have been more pragmatic to their own national interests (Azerbaijan, Ukraine, Kazakhstan, Uzbekistan, and Turkmenistan).[11] This increased Russian influence has not come without cost, as regaining the sphere of influence once taken for granted as the Soviet Union is taking a toll on a constrained Russia that does not have the resources it once had during the Cold War.

Out of the peaceful, yet economically tumultuous, dissolution of the Soviet Union came the revitalized Russian state. The ascension of Vladimir Putin to the presidency in 2000 coincided with several economic reforms, specifically in its enormous energy export sector. These reforms led to a stabilized Russian economy, an economy based upon high energy prices that leaves Russia vulnerable. Putin's popularity with the Russian people was the result of this good fortune. Although Russia had brought itself back from the brink of becoming a minor power to a regional hegemon, the traditional notions of power it enjoyed as the USSR could no longer be sustained. Russia no longer had the resources to maintain a large navy and air force and was stopped cold in Chechnya by underequipped rebels in the 1994–96 civil war. Russia of the 21st century, also known as "Putin's Russia," is not the military powerhouse that the Soviet Union was, and the reach of this new Russia is limited. Its population has been cut in half due to the Soviet breakup.[12]

This population is also in decline due to a combination of emigration, birth mortality rates, and life expectancy. The life expectancy of Russian men is well below the expectancy of the states of the West. Russian men's expected life span is a dismal 60.1 (compared to 73.2 for Russian women).[13] Sixty percent of Russian men smoke, each Russian on average consumes nearly four gallons of alcohol per year (compared to around two gallons for the United States), and half the population is overweight.[14] Russia's military has therefore had difficulties in drafting capable men. It has faced even more challenges since trying to switch from drafted to voluntary forces. Most importantly, this population decline has severely impacted the pool of military-eligible men. All these factors are challenges to Russia's ability to project power externally and push the state to focus on new forms of power projection.

In this volume, we focus on how the new forms of power are directed by an examination of Russia's use of coercive diplomacy. What motivates power politics actions, and what are the consequences of Russian foreign policy choices for its neighbors, the United States, and even Russia itself? How does Russia use the power it does have and when; and also when does it choose the accomodationist path? When does Russia choose to cooperate with its neighbors and other international powers?

Foreign policy scholars often fail to examine both the sources of action and the success of foreign policy objectives in light of goals. They also fail to truly examine the nuanced contexts of foreign policy action. Blanket theories of foreign policy processes are empty without an examination of the linkage between domestic demands and opinions, on one hand, and external options and constraints, on the other. This two-level examination of foreign policy should be the basic foundation of any foreign policy theory.[15] Theories of foreign policy also fail if they do not examine what might be termed historical animosity in the form of rivalry.[16] These background conditions, along with the issues involved, the region under consideration, and the strategic choices available to states help determine the foreign policy path for states with a moderate amount of power.

This survey examines Russia and its continuing, but shifting, rivalry with the United States, Russia's actions in cyberspace, its newfound power in its energy endowments and the use of energy power as political leverage, and Russian engagement of other great powers in the Arctic in order to understand how power is used in its modern forms. With such a broad sweep of Russian foreign policy actions and considerations, we hope to be comprehensive in our examination of Russian paths towards diplomacy or conflict in the early 21st century.

In many ways, this book is an application of a theory rather than an examination of Russia itself. We are not Russian scholars but international relations generalists looking to extend and examine our theories about how new forms of power fit into the traditional power politics framework. Our theory of situational coercive diplomacy takes many factors into account and can also be used to understand outcomes. Only time will tell if our theory is correct in its predictive power, but for now we can be assured that our framework explains the current challenges and operations of the Russian state. The question is how long this will hold and if our theory applies to other midlevel actors in the international system.

We take no normative stance on Russia's choices but only hope to examine the effectiveness of these choices in context. Foreign policy is much more than expressing a state's interests; it is about achieving success in the international realm according to the values of a state and the power at hand. On this point, Russia has had frequent missteps, since its use of coercive diplomacy often results in failure. We note in this volume that a better path might be exhibited in the Arctic, where institutions and international law restrict Russian ambitions but also encourage the state to work with other actors in the region to come up with viable solutions to international considerations. It is hoped that this volume can help others present a model that explains the foreign policy choices of powerful states who may seek to act as a regional hegemon.

Russia as a Geopolitical Enemy

Sometimes people are right for the wrong reasons. One such occasion occurred during the run-up to the 2012 presidential election in the United States. Republican nominee Mitt Romney remarked, "I'm saying in terms of a geopolitical opponent, the nation that lines up with the world's worst actors, of course the greatest threat that the world faces is a nuclear Iran, and nuclear North Korea is already troubling enough, but when these terrible actors pursue their course in the world and we go to the United Nations looking for ways to stop them, when [Syrian President] Assad, for instance, is murdering his own people, we go to the United Nations and who is it that always stands up for the world's worst actors? It is always Russia, typically with China alongside, and so in terms of a geopolitical foe, a nation that's on the Security Council, that has the heft of the Security Council, and is of course a massive security power—Russia is the geopolitical foe."[17]

Newspapers, pundits, and the Internet were in an uproar. President Barack Obama termed this line of thinking a "Cold-War Mindwarp."[18] Even Romney's supporters were mystified. How could a presidential candidate in 2012 still have a Cold War mentality? The problem for many who reacted so negatively is that Romney was right; Russia is currently a foe of the United States, but this view comes with several caveats. For one, Russia is not more troubling than Iran or North Korea. More importantly, Russia is not a consistent enemy nor a threat to American power.

Russia is not a "geopolitical" threat. It is a regional power that challenges the United States in its sphere of influence, which is post-Soviet space. As the *Washington Post* notes in its fact-check article of Romney's comments, Russia has not always challenged the US and the Security Council to protect "bad" actors.[19] It was supportive of action against Iraq, Afghanistan, and Libya. Russia continues to support the UN's opposition of Iranian and North Korean nuclear ambitions. Where it actively and consistently checks American power is in post-Soviet space and in locations where it has deep political ties. South Ossetia and Abkhazia, the separatist enclaves in pro-West Georgia, come to mind as examples of this process. Russia does all it can to protect those who have close ties to the state and has even granted citizenship to anyone with Russian ethnic ties, no matter how far outside its borders they might be.[20]

At other times, Russia may support actors beyond its regional sphere of influence, but it only does so when there are background ties that predate the reemergence of the Russian state. As one of Russia's last major allies remaining from the Cold War in the Middle East, Syria is one such state. In the same way the United States defends Israel, Russia defends Syria because of historic ties and the fear that the loss of this connection will mean the loss of prestige and an important foothold in the Middle East.

Romney doubled down on his comments with an opinion piece in *Foreign Policy*. "Across the board, it [Russia] has been a thorn in our side on questions vital to America's national security. For three years, the sum total of President Obama's policy toward Russia has been: 'We give, Russia gets.'"[21] Unfortunately, this idea is also wrong; Russia has not been rewarded for its actions. In fact, it has often failed to achieve its foreign policy objectives, and this is the point of our story. Over and over again, when Russia pushes, coerces, and bullies, states do the opposite of what it wants. The United States is trying to work with Russia to establish a new set of relations (the reset button that seems to have fallen out of favor as an idea), but as we document, the two states are still mired in a regional-level rivalry over questions of post-Soviet space. Compromise is a part of the normal course of foreign policy, and to ignore this is folly. By trying to accommodate Russian demands, the Obama Administration has been trying to mollify Russian concerns and potential aggression. It is also trying to change the "hearts and minds" of ordinary Russians, who still vilify the United States as if the Cold War never ended (see, for example, Table 3.2 in Chapter 3). Terminating a rivalry can be a long process; part of that process is to change the opinions of the public in a rivalry state. The current path seems to be the best option available to achieve a plethora of goals that both countries share.

In this book, we will outline why exactly Russia is still the main rival of the United States, but here it is important to put that statement into context and discuss what exactly it means to be a rival and what responses are in order. The reason Russia is still the number-one foe of the United States is because no other threat has materialized, because interests clash in post-Soviet space, and because public attitudes in each country have not changed with shifting geopolitical realities. China is currently not yet a rival of the United States; it is simply a competitor, and often a competitor with which the United States works closely.

The greatest threat to American power in 2014 does not materialize from states but from situations like economic decay and globalization, climate change, failed states, or ethnic conflicts from immigrants assimilating into countries. That Russia is a rival just means that interests between the two states will clash. While these clashes might result in regional violence, Russia is clearly not an existential threat to any state outside of the post-Soviet sphere. There is nothing to fear from Russia, and as it continues to develop beyond its Soviet legacy, it continues to find that coercive diplomacy fails and that other options are more successful in achieving political goals. For example, Russia has fully joined the world economy with its ascension to the World Trade Organization in 2012. Romney was right to say that Russia is a foe and that it can be dangerous, but this danger is localized to one region and should not be a motivating factor to increase weapons production or gather allies in the territories of the former Soviet Union. Instead, Russia's status as a reemerging giant using new forms of power just means that Russia cannot be ignored. Its interests are stable and well known; ignoring Russian preferences only sets the stage for diplomatic battles, as we have seen in such areas as Iraq, Iran, Kosovo, and Syria.

Russia can and will be integrated into the world system; the question is how and what form this integration will take, because if it is painful and difficult, the repercussions could last for generations. Status matters, and status inconsistency—or not being given the proper place due to a state such as Russia—results in insecurity and the need to demonstrate the state's worth on the international stage.[22] Conflict can be managed and avoided, but only by moving away from the Cold War mentality that many people still have. The Cold War is over, and capitalism and democracy "won," but that was never really the reason the Cold War was so dangerous in the first place. The Cold War was dangerous because it set both powers up for a battle of influence. This battle continues, but the battlefield and stakes have changed.

Russia's Place in the World: Time for a Reevaluation Considering New Forms of Power

The collapse of the Soviet Union threw Russia into a decade of economic and political ups and downs, where eventually the economy and government were stabilized, yet this turmoil of the 1990s in Russia did not create an environment for democratic institutional development.[23] Instead of creating and stabilizing institutions that would allow for a prosperous democracy, the Russian political elite auctioned off state resources in exchange for political stability through the economic elite.[24] Russia's state-building capacity was crippled, institutional evolution was stalled, and the country became the corrupt, oligarchic system that the West was trying to prevent. Power was in the hands of the oligarchs, who set up institutions salient to their interests. Some had all of the power, but most did not. This incredible asymmetry of power in 1990s Russia allowed for its unique institutional development, a development which did not have state building in mind.[25] Old institutions that were not ready for the transition to democracy, along with a deepening economic crisis, allowed for this institutional path to be taken.

In 1998, the bottom of Russia's struggling market economy fell out. The bullish privatization of Russia finally ran out of steam. Although Soviet property and industry was being sold off, no investments were actually made in growing the new market economy.[26] Along with the reduction in crude oil output as well as the global price of oil, the slow rise in the Russian GDP led to the 1998 economic collapse. Furthermore, the Russian government began bouncing checks on government bonds, and international investors soon realized that the government was bankrupt and that Russia was no longer a sound investment. Inflation of the ruble led to hyperinflation, leaving most of the Russian people destitute.[27] Yeltsin deflected blame from himself and appointed and removed three prime ministers within about a year's time period, finally settling on the relatively unknown former KGB officer Vladimir Putin in August 1999. Yeltsin then resigned in December 1999, citing deteriorating health conditions, and Putin became acting president.

Putin quickly galvanized public support after an act of terrorism on apartment buildings in the outskirts of Moscow, and Chechen nationalists were Putin's prime suspects.[28] He then promptly sent Russian troops into the troubled Caucasus region and won quick victories, which almost overnight made him a Russian hero after repeated failures in the area in the decade prior. Putin won the 2000 presidential election in a landslide, and the era of "Putinism," or economic and political reform in Russia done Putin's way, had begun.[29] This reform saw most power consolidated within the office of the presidency and allowed for an unprecedented economic recovery unmatched in Russian history. Putin

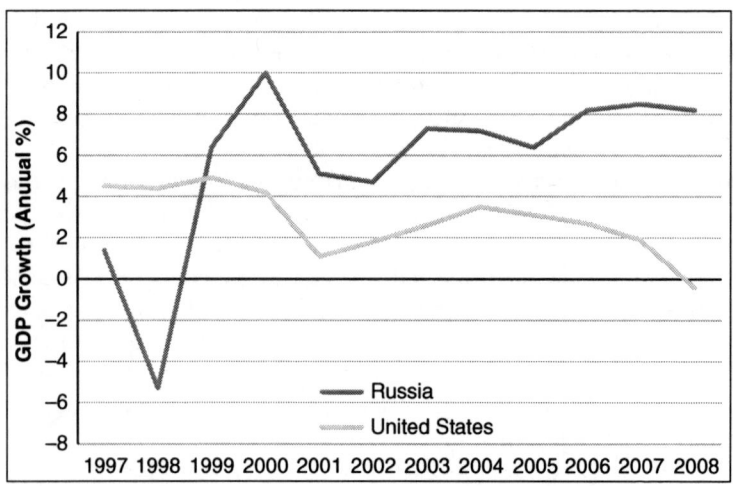

Figure 1.1 Russian and US GDP growth comparison 1997–2008

Source of data: *The World Bank*. Accessed 9/14/2012. http://data.worldbank.org/indicator/NY.GDP.MKTP.KD.ZG

was able to stabilize the Russian economy, and Russian GDP growth outpaced many of the economic powers of the West, including the United States, until the collapse of the global economy in 2008 (see Figure 1.1). All of these events then led to more confidence in the Russian foreign policy regime, which in turn set up the rise of Russian coercive diplomacy under a new context. This new context is now found in Russia's energy, cyber, and maritime power.

Russian Energy Power

Where Russia does have the ability to get other actors to do something they otherwise would not do is through its coercive energy policy. Russia has natural gas and pipeline leverage over most states of post-Soviet space, either in the form of gas dependency or pipeline access. For example, Kazakhstan needs Russian pipelines to get its natural gas exports to world markets, and we argue that Russia takes advantage of this. Similarly, Belarus is 100 percent dependent on Russian energy imports, and Russia also takes advantage of this. When Putin ascended to the presidency in 2000, reforming the Russian energy sector was on the top of his priority list.

Putin's first reform effort was to remove the oligarchs who could challenge him politically and replace them with his own political allies, many of whom were either former KGB associates or political upstarts from Putin's hometown of St. Petersburg.[30] It was especially important to the new Putin Administration that Russia's natural resource companies be controlled by friendly faces, and to do this, those who did not share Putin's vision for Russia were controversially removed. Rem Vyakhirev of Gazprom was removed as chairman of the natural gas conglomerate. Media tycoon Vladimir Gusinsky of Media Most was jailed on charges of embezzlement. Sibneft's Boris Berezovsky was threatened with jail, yielded his interests in the oil company, and fled to the United Kingdom. Russian Central Bank's Viktor Gerashchenko was removed by the new president for allegedly siphoning off state resources. Finally, the richest and most vehement opponent of the new Putin Administration, Mikhail Khodorkovsky of Yukos oil company, had his company seized by the Russian government and was jailed on charges of forgery, tax evasion, grand theft, fraud, extortion, and embezzlement.[31]

With most of his potential political threats or challengers either in exile or jail, Putin moved to either bulk up or consolidate the oil and natural gas industries into the state-controlled giants Rosneft (oil), Transneft (oil pipelines), and Gazprom (natural gas and natural gas pipelines).[32] These conglomerates bought up media outlets, Internet companies, and the like. Putin had consolidated his power, nationalized Russian natural

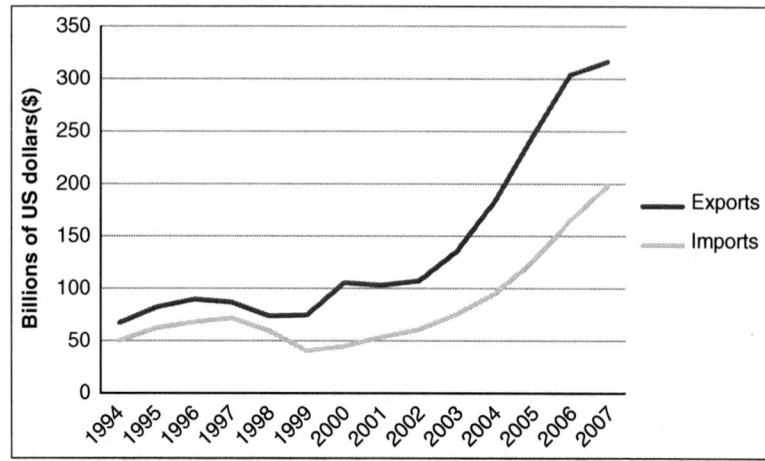

Figure 1.2 Russian exports and imports 1994–2007

Source of data: Marshall I. Goldman. 2008. *Petrostate: Putin, Power, and the New Russia* (Oxford: Oxford University Press): 92.

resources, and put his friends in positions of power so that his control of Russia's economy was safe. With these moves, as political freedoms dwindled, economic prosperity boomed.

The first few years of Putin brought Russia into the world market. Figure 1.2 shows the impressive rise in both imports and exports, which began to quickly put Russia back into the realm of the world's great powers. In 2000, nearly 50 percent of Russian exports were comprised of oil and natural gas, and by 2007, this figure reached near 67 percent.[33] Russia has vast reserves of oil and natural gas, and with the revival of the Russian economy and the rise in exports, Russia had newfound power. Under the Soviet system, Moscow was unable to harness its natural resource potential due mainly to the setbacks a command economy inherently caused.[34] Once integrated into the world market, Russia had wealthy and demanding energy customers in Europe and Asia. This resource wealth meant that Russia was able to pay off its foreign debts, gain leverage on many states in post-Soviet space in the form of energy coercion, and begin to project its power in the international arena.[35]

Russian Cyber Power

Russian cyber power is considered to be one of the savviest and technologically advanced in the world. Along with China and the United States, Russia is considered one of the "heavyweights" in its abilities to deface websites, steal information, and cripple states' infrastructure via the web.[36] Furthermore, Russian cyber defenses are also some of the best in the world, with the Federal Security Service (Federal'naya sluzhba bezopasnosti), or FSB, the Russian organization that is the successor to the more well-known KGB during Soviet times, filtering most Internet traffic in and out of Russian borders through its servers. It also boasts of being the home of Kapersky Labs, the computer security firm that helped discover the Stuxnet worm and the Flame virus.

Russia brought international attention to itself when it was blamed for launching a series of offensive cyber incidents upon the state of Estonia as a reaction to the government in Tallinn's relocation of a World War II Soviet-era grave marker. Furthermore, Russia is the only state thus far that has launched cyber operations against a state concurrent with a conventional attack when it defaced and shut down various Georgian government and private websites during the five-day skirmish it had with the Caucasus country.

Russia has broken into several American-based networks—moves that continue to exacerbate the ongoing rivalry with its Cold War adversary.[37] However, Russian cyber power has been restrained up to this point, where the potential of Russian capabilities has not matched its actions in cyberspace. Chapter 4 uncovers why this is the case. We ask how Russia uses its cyber power and also why it has been restrained to this point. What is the future of cyber power, and do the means achieve the ends?

Russian Maritime (Arctic) Power

Another new context of Russian power lies in its shifting use of maritime power. The Russian navy, which once had global reach and was able to counter American ships anywhere in the world, has been drawn back to Russian coastal waters, as the only Russian naval base outside of immediate proximity is in nearby Syria. The Soviet navy had a global reach for its global interests; the 21st century Russian navy is being specialized to protect Russia's regional interests, especially in the Arctic region. Russia has the most to gain out of climate change and an ice-free Arctic Ocean. Russia will be able to access oil and natural gas off its continental shelf previously untapped because of the logistical problem that comes with a frozen surface, subsequently helping increase its energy power; Russian ports bordering the Arctic Ocean will become important seaports for a new northern sea route for international trade; and Russian Arctic coastal waters will now be accessible to Russian fishermen. An ice-free Arctic, therefore, requires a Russian navy capable of defending Russia's interests.

Figure 1.3 demonstrates that Russia has spent much of its newfound wealth rebuilding the Russian military-industrial complex. One of Putin's main priorities is the establishment of Russian prestige and respect internationally, and a big part of this is the reestablishment of Russian military power through the revitalization of the navy.[38] Figure 1.3 indicates that after 2000 and the ascendency of Putin, Russian military expenditures, in terms of yearly percentage change, skyrocketed. Russia has modernized its navy and army as well as reestablished its coveted arms sales to former allies of the Soviet Union, such as Cuba and Syria. These states are adversaries of the United States, and American policymakers have seen this resurgence in arms transfers between Russia and American opponents as a threat to international stability.

The partial restoration of Russian military power and the unprecedented economic growth have brought Russia back into the club of midrange powers such as Great Britain, China, Brazil, France, and India. However, Russia does not have the power to extend its arm globally, as the Soviet Union once could. Russia's economic power is largely constrained to former Soviet space as well as parts of Europe. Russia is outmatched economically and militarily by the United States; therefore, disagreements with its Cold War adversary are largely limited to areas where Russia can reach and apply pressure. These areas are usually in the former republics of the Soviet Union, and rarely do Russia and the US have confrontations outside of this region.

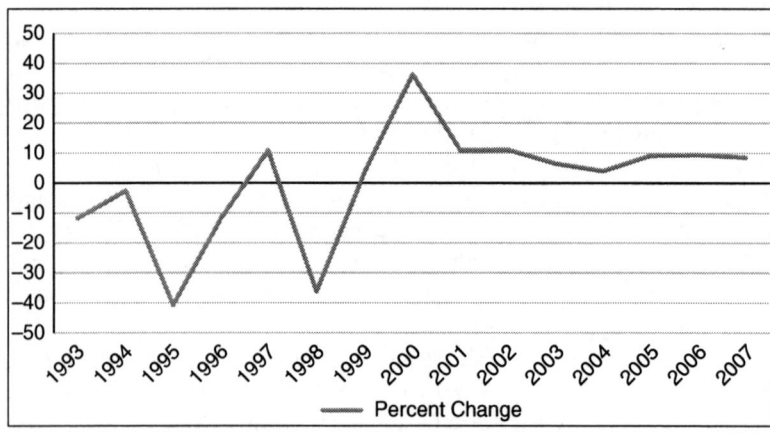

Figure 1.3 Yearly percentage change in Russian military expenditures
Source of data: SIPRI Website: http://milexdata.sipri.org/

For now, Russian use of coercive power is limited. However, throughout the former Soviet space, Russia does have the ability to get other actors to do something they otherwise would not do, and these states bear the brunt of its coercive diplomatic might. The research in this book uncovers just how Russia uses coercive diplomacy and power politics in its sphere of influence, and how the West, particularly the United States, reacts to Russian foreign policies. This is important because the US–Russian rivalry of old is being rekindled in post-Soviet space, as Russia has held steadfast in what it considers its exclusive sphere of influence. Many American incursions and influences in Russian space have been largely not tolerated by the Putin Administration.[39] However, there are examples of US–Russian cooperation in other parts of the world, and perhaps policymakers on both sides can learn from these examples of diplomatic solutions to gather lessons about the limits of coercive diplomacy.

Another arena where Russia is able to counter US and Western interests is in the UN Security Council, where Russia still has a permanent seat and unilateral veto power. Along with China, Russia has caused divides in resolutions introduced in the Council, complicating Western and US foreign policy goals globally. Examples of late include the civil unrest in Syria and opposition to NATO interventions in the Balkan and Middle Eastern regions. Preventing the foreign policy goals of the United States is the behavior of a rival, a situation that we argue in this book is alive and well and did not end with the death of the Soviet Union. Russia, therefore, is back on its old path but not with the scale and reach that it had in the past.

Situational Coercive Diplomacy

There is no blanket foreign policy theory that can explain Russian foreign policy choices, particularly under the context of new forms of power. Interests are constructed by identity, values, and ideas held by the Russian elites and population.[40] This nuanced and constructivist view of Russian foreign policy is missing in the literature, and here we offer a path towards uncovering how Russia uses traditional and non-traditional forms of power through motivations that go beyond the national interest.

Russia acts in a way that may preserve and display its strength in the region to reassert its traditional control but also in a way that is not typical of the prudential-measured restraint advised of great powers.[41] It frequently oversteps its bounds and is checked by Europe or the

United States. Its actions in Georgia during 2008 are a prime example of this. In this situation, Russia acted according to its values and the identity that it had dominance in the region, and it discounted the protection the United States would offer its ally Georgia, a view that happened to be correct.[42] Unfortunately for Russia, international condemnation resulted from these actions, and further actions would have generated economic reprisals.

Theories of foreign policy cannot disconnect the domestic sources of foreign policy from the diplomatic interactions that take place between states. Russia engaged Georgia because it had the power to do so, because that action was part of its identity as great power and regional hegemon, and because the public was in favor of such action. Scholars often leave the public and elite sources of opinion out of the story of why foreign policy actions are considered. When these factors are considered, scholars then often leave out the conflict dynamics inherent in the dyadic processes of states. We hope to close these gaps here. This is not to say that public opinion dominates or shapes foreign policy, but it can be an important factor that should not be overlooked.

The first step is to recognize that it is not the system that matters but the regional subsystem. Utilizing the work of Buzan, Nygren calls these subsystems regional security complexes, where Russia is competing with other powers for influence in three sub-complexes of post-Soviet space: the European, Caucasian, and Central Asian.[43] The broader context of Russian geopolitical constraints is not relevant because Russia is no longer a geopolitical actor. It is an actor mainly concerned with regional power dynamics rather than broader ambitions of global hegemony. Despite diplomatic grandstanding where it may challenge US power, Russia cannot hope to challenge US might, or even European might, on most geopolitical questions. It can, however, assert its regional might in post-Soviet space. We therefore call our theory of Russian foreign policy action "situational coercive diplomacy." Russia's use of coercive diplomacy is shaped by its rivalries, domestic public opinion, the salience of the issues under consideration, and the region where action is being considered. It is a product of its environment, history, and values. Its use of power is also constrained or expanded by the domain in question. In the form of energy power, Russia has near unlimited abilities to flex its might given the nature of this power. Russian cyber power is much more restricted, and this is evident by its clear restraint in Ukraine in 2014. This is the topic for discussion of Chapter 4.

The first consideration is historical animosity. Russia acts under the weight of history each time it engages in foreign policy practices. Its past history—mainly including rivalries but also other various historic slights—shapes its values as a country and therefore its interests. Each time it comes into contact with another foreign actor, it is best to ask what sort of current relationship it has with that country and also what sort of past relationship it has had.

The next consideration is domestic public opinion, including both elites and the general public. Some theories suggest that public opinion on foreign policy matters is shaped by the leadership of a state.[44] This is true to some extent, since by and large the collective will of the people is tough to modify and shape in a consistent manner. It is easier for the leadership to take advantage of current moods and feelings rather than to shape them. Therefore, public opinion of elites cannot be ignored. But in this regard, we must also examine popular opinion to understand how elites shape options or how expressions of dissatisfaction can be utilized by the leadership to achieve political ends.

The salience of the dispute under examination is the next critical step in evaluating Russian foreign policy. The types of disputes under contention, plus the collective sense of importance about various disputes, also matter a great deal in determining foreign policy action. "The salience of an issue is the degree of importance attached to that issue by the actors involved."[45] Disputes that are more salient on average have a higher likelihood of escalating to the use of force. The most salient issues between states tend to be territorial in nature. Ignoring the value placed on issues misses a great deal of predictive information that can be leveraged when explaining foreign policy action. The interesting paradox is that disputes of a transcendent quality tend to be tougher to settle and more likely to elicit force than disputes that have concrete stakes. Concrete issues are fairly easy to settle since they are divisible, but intangible disputes are more difficult to settle since the value placed on them tends to be priceless in some ways. Although territorial disputes are seemingly concrete issues in that territory is divisible and physical, they are usually intangible due to nationalist or symbolic feelings in the public and elite spheres. The stakes become so high that it is nearly impossible to settle the issue in favor of both sides. Understanding when, where, and which issues are salient to Russia will help determine which events are likely to see an escalation of violence.

The region of interest is also important for determining foreign policy action. Feklyunina argues that Russian identity is inextricably tied to control of its former empire, thus post-Soviet space is essential to what Russia is as a modern state and how Russia sees itself.[46] Some states have a global portfolio and a global reach, while others are more limited in their capabilities. Much to the dismay of Russians, Russia is a limited state in terms of capabilities. It took until the mid-2000s for Russia to even be able to settle pressing issues inside its own borders (for example, Chechnya). Russia is now able to assert its power in its former empire, but beyond this area it is limited in its abilities to project power. This may change in the future, but as of now it is prudent to ask where the conflict is located in order to understand how much Russia can do to affect the situation.

Various factors interact to shape Russia's foreign policy outlook. Foreign policy is not made or developed through the input of one variable but through the interaction of many influences. This study takes this perspective into account and therefore attempts to apply, predict, and understand what factors have motivated Russian foreign policy action and in which direction.

The Plan of the Book

The plan of the book is to uncover the issues important to Russia's foreign policy goals and the tactics it employs to deal with them. It is an exploration of Russia's use of new coercive power and the failures of its application. The next chapter details the framework and overall theory we utilize to guide our research. Our theory of situational coercive diplomacy uses factors including issue salience, rivalry presence, region, timing, and public opinion to uncover why Russia interacts with other states the way it does. Why is Russia more cooperative with certain states than others? Which factors lead Russia towards a path to war? We argue that it all comes down to the issues under dispute, particularly territorial and energy issues in post-Soviet space. Chapter 2 outlines the tools to be used and the overall theory of the book.

Chapter 3 argues that the United States and Russia are still principal rivals and that the rivalry relationship did not terminate with the collapse of the Soviet Union. Rather the rivalry continues and has shifted from a geopolitical rivalry for global influence to a regional rivalry over disagreements in Russia's backyard. Issues under dispute include alliance systems, arms transfers and missile systems, and issues involving post-Soviet space. Chapter 3 also tests our public opinion measure of issue salience. We present Russian and US public opinion of each other's states to show that domestic sources of rivalry can serve as a tool for measuring the continuation or termination of a rivalry. We then present Russian public opinion polls about the international issues that Russia has found pertinent to its foreign policy objectives. We find that public opinion is a relevant measure of issue salience and can help predict Russia's foreign policy tactics for each issue, but public opinion alone will not drive action. It often reacts after the situation and serves to funnel the will of elites.

Chapter 4 presents and compares the power Russia has in cyberspace, discusses when it uses coercive tactics there, and examines how this compares to other states. We will discuss how cyber conflict is conceived and analyzed in popular and academic discourse, followed by a section uncovering who is behind Russian cyber capabilities. Then we will show how the capabilities Russia contains compare to other cyber "heavyweight" states that have advanced technological capabilities, followed by an analysis of how Russia has used this technology as a foreign policy tool against its post-Soviet adversaries of Estonia and Georgia, as well as its long-time rival, the United States. We also examine Russia's use of cyber tactics in the context of its involvement in the destabilization of Ukraine in 2014 in the concluding section. As the face of diplomacy changes with the advent of cyber capabilities, it is important to understand how this new tool is used and when, if at all. We find that cyber incidents and disputes are a limited option Russia uses, in that it has the power to do much more but has been constrained by norms and international condemnation in its actions in the cyber realm. This is a puzzling finding in that conventional wisdom paints Russia as a cyber aggressor rather than a restrained power.

Chapter 5 is an examination of Russia's use of coercive energy policy. We use certain issue-based and contextual variables to see if these factors have an effect on the amount Russia charges these states for its natural gas. Russia charges its post-Soviet customers somewhere in between market prices for the EU and the heavily subsidized domestic price range. Here we argue that factors such as close ties to the Russian government, ties to the West, whether or not a militarized interstate dispute (MID) has occurred between pairs of states, whether or not the country is a pipeline transport territory for Russian energy exports, and the amount of ethnic Russians living in the CIS are all relevant to the examination of Russia's use of energy as a weapon towards its post-Soviet customers. Several panel regressions from the years 2000 to 2011 are employed for this model.

Chapter 6 continues our examination of Russian coercive energy policy and applies the theory of situational coercive diplomacy to Russia's use of its energy policy and its measurements of issue salience, rivalry presence, the particular region under examination, and public opinion to uncover the importance of energy to the Russian state. It also uses constructivist literature on Russian self and external identity to further argue for these issue-based motives of coercive energy policy usage.[47] Qualitative techniques are used to uncover in depth the gas pipeline politics Russia uses in two important regions of post-Soviet space: Central Asia and the Caucasus. Through the framework of the theory, we find that US involvement in energy issues of post-Soviet space will lead to more coercion, whereas another major power, China, does not force Russia into a coercive energy policy strategy for Central Asia.

In Chapter 7, we look at Russia and its relations with other states in the Arctic area in a unique examination of when coercive diplomacy is not used, and accommodation and cooperation is apparent with Russian foreign policy action. We find this option is not taken because the conditions laid out in our theory are not in operation. Along with the other coast-bordering states of the Arctic Ocean: the United States, Norway, Canada, and Denmark, Russia has important future stakes in this remote region of the world. Natural resource exploration and extraction, opening shipping lanes, and fishing waters are all high-stakes issues about which these states are involved in territorial disputes. However, as Russia can rightfully claim nearly half of the underwater territory of the Arctic Ocean according to the laws of the sea set out by the United Nations, it sees no reason to get into coercive disputes with its Arctic neighbors.[48] Why fight when it can get what it wants and maybe more by cooperating through international institutions? The findings of this chapter are surprising in that cooperation is found and paves the path for analyzing Russian foreign policy, which is the topic of the final and concluding chapter in this volume.

Overall, in this volume we examine, predict, and uncover the sources of Russian foreign policy actions after the Cold War. Our work demonstrates that Russia has taken a new path, and a new lens must be applied to study this process not only for Russia, but for all midrange powers in the contemporary era. We hope the tools, lens, and theories applied to examine Russian foreign policy can also help explain how foreign policy is used by other actors.

2
Russia's Foreign Policy Choices and the Application of Situational Coercive Diplomacy

Introduction

Russian foreign policy in the modern era is based on traditional notions of strength and power determining actions and motives in the international realm. Even in an age when Russian power projection capabilities have been degraded, the goal remains to utilize coercive foreign policy means to achieve objectives with new forms of power.[1] Nevertheless, Russian foreign policy dynamics are more complicated than simple power-maximizing behavior. Russian foreign policy is based on coercive diplomacy in a regional sphere of influence, according to certain contexts that motivate action or perceptions. Russia's use of power politics tactics is limited to instances where it has a long-standing rivalry, the public supports such action, and when there are salient foreign policy issues at stake. The location or region where a foreign policy challenge arises is also an important context. Much like the American (Monroe) doctrine of regional control in the Western Hemisphere, Russia operates similarly in the Near Abroad region.[2] In this chapter, we unpack these processes and suggest a framework to examine Russian security interests after the end of the Cold War. Russian power is on the rise, but this is a different power in that international constraints and the current security environment will hinder Russia's ability to have a global influence on the actions of other states. This pushes Russia to utilize new forms of power such as cyber, energy, and maritime claims and legal power where they have not in the past.

Foreign policy is the outputs of a state's national security program. It is the goals and objectives of a state as it operates in the international realm. If foreign policy is the output of a state's foreign policy apparatus, what exactly are the inputs? Moving behind the internal decision-making apparatus of a state, what contexts make foreign policy relevant and push it to the top of a state's agenda? This question seems to have been left out of the contemporary foreign policy discourse. Evidence we marshal in this volume leads us to argue that the foreign policy choices states make result from interactions between domestic prerogatives, including the salience of various issues at stake, the public opinion preferences of elites, and international contexts. International contexts generally take the form of past conflict interactions, or rivalries, and the regional dynamics in which a state operates.

As Vasquez notes in his examination of foreign policy studies, "even the simplest review of policy that compares intentions with consequences and sees whether means actually achieve goals has been left undone."[3] In this volume, we hope to examine the goals and intentions that lead to action and then to evaluate the outcomes, if available, of Russia's foreign policy actions. We note that power politics using unconventional strategies are the foundation of Russian external decision making, but this finding is not universal. When actions are taken under certain contexts, foreign policy outputs take the form of power politics strategies, which typically rely on the use of force, escalating bargaining tactics, and displays of power to change or moderate behavior in a target. Here we examine Russia's use of unconventional power strategies and seek to uncover the outcomes of these tactics in relationship to the goals.

Context and situations matter a great deal, so do perceptions and the elite plus public sentiments. As we will demonstrate in this book, Russia's reliance on new power politics tactics has not increased its reach or power; quite the opposite. Russia's use of power politics tactics has only hindered its ability to achieve its foreign policy goals. Simply put, coercive diplomacy often fails.

Moldova is a perfect illustrative example of this process that indicates the dark side of these new power politics practices. Russia has pushed Moldova to bend to government-controlled Gazprom in energy matters. Aslund Anders, writing an editorial for the *Moscow Times* notes, "Although Moldova has given up control and ownership of its gas pipelines to Gazprom, it has received nothing in return. Gazprom still demands exorbitantly high gas prices of more than $400 per 1,000 cubic meters, far more than it charges EU customers. The obvious lesson is that any concession to Russia will be punished by Russia with further demands."[4] There is always a reaction to power politics actions, and the shape of this reaction determines the course of international relations.

Using coercive foreign policy tactics may work on a few occasions, but on balance it is a prescription for failure, no matter the shape of the tactic. We will document this process in this book. The use of coercive foreign policy is the hallmark of Russian foreign policy, and when these tactics are used, Russia ends up driving its target away from the Russian sphere of influence and more towards the West. The West is a willing partner with many post-Soviet states since various actors, such as the United States, the European Union, and the North Atlantic Treaty Organization (NATO), seek to expand their own institutions to new regions to maximize their interests.

Russia has particular foreign policy goals, but it tends to fail in the achievement of objectives when power politics practices are used repeatedly. Anders notes, "Russia's policy in the post-Soviet space is costly to all, but most of all to Russia because it gives the CIS countries little choice but to turn their backs on Russia. Thus there is little surprise that Moldova does whatever it can to get closer to the EU."[5] It has secured loans from the EU and has also made inroads towards trade agreements with the organization. Often, power politics moves achieve the complete opposite of what was intended. This is the true dark side of the power politics path. These options may seem necessary and universal, but in fact, they often fail to achieve the goals set by the initiators, which then necessitates, in the eyes of policymakers, further power politics responses and the use of force. The conflict spiral starts here, with the overuse of power politics strategies.

When Russia chooses the opposing path to coercive diplomacy, or a more diplomatic and accommodationist path, Russia's foreign policy ends are more likely to be achieved. The question then is what is the main reason Russia would choose to utilize coercive foreign policy tactics? We argue that there are two determinants for this choice: the type of issue at stake, including the salience for this issue within Russia and among foreign policy elites, and the international context, including the level of rivalry between Russia and the contending side, plus the region in which the foreign policy question arises. The outcome of these factors determines Russian foreign policy strategies directed towards the states that operate in the region. There is no consistent theoretical tradition currently offered that can determine outcomes based on these inputs; therefore, we term this theory "situational coercive diplomacy." Here we offer a new framework to determine which choices states with at least a moderate level of power will make in the realm of foreign policy.[6] This volume is about Russia, but the theory can be applied to states in similar circumstances. Raw calculations of power devoid of context do not provide a prescription for consistent analysis and prediction. We hope to bridge this research and policy gap with the framework offered here and support our predictions with a series of examinations of current Russian foreign policy practices.

In this book, we ask three major questions. The first is how is Russia using its foreign policy abilities? Is it using the power politics–coercive diplomacy path or choosing the more accommodationist, compromise-based system of foreign policy management? The second question is what foreign policy contexts are present when Russia makes the choice between coercive diplomacy, on the one hand, and accommodation, on the other? The final question is what is the outcome of these foreign policy choices? Did Russia achieve its strategic objectives, or does the choice of tactics lead to rivalry and competition? If coercive foreign policy objectives failed, then why use the tactic? The rest of this chapter will outline our framework for analysis. The choice between power politics and accommodation is the key foreign policy choice states make, and this book examines these choices, the reasons they are made, and the consequences of these actions with Russia as the focusing state.

Current Views of Russian Foreign Policy

There has been a recent surge of interest in Russian foreign policy. Witnessing a transformation in how power is used and compiled by a fallen and then rising power is a rare event that spurs ideas and research. Few go as far as Lucas in asserting that there is a "New Cold War."[7] This revived Cold War is being fought with cash, resources, and propaganda to expand Russia's influence throughout the globe, according to Lucas. Money is the motivating factor for Russian action since energy prices have skyrocketed since 2001, and this has given Russia the latitude to once again operate in the international sphere. Kubicek makes a similar point, but less forcibly.[8] Russia now has the ability to protect its energy interests in post-Soviet space after falling behind prior to 2001. Other states, such as China and the United States, took advantage of Russian weakness and sought to acquire influence in the post-Soviet region. Rising commodities prices allow Russia to be a revisionist power and seek to once again become a great power.[9] This is particularly true of Russia's revitalized interest in Ukraine.[10]

Literature in the foreign policy analysis field takes historical and behavioral approaches to explain contemporary Russian international interactions. Sherr and Averre both look at Russian behavior through a historical perspective to explain the growing contemporary discord between post-Soviet Russia and the West.[11] It is also found that although the EU and Russia have shared economic and security interests, their differing diplomatic styles have led to an unexpected level of discordant relations.[12]

Allison and Ambrosio concentrate on how Russia decides to intervene militarily in post-Soviet space and when it is more diplomatic and accommodative.[13] The growing authoritarianism under Putin and his backlash towards democratic trends in post-Soviet space serve as examples of the reasons behind the more coercive diplomatic strategies since the turn of the century. Another important purpose of these strategies is to rein in American power and tip the balance eastward towards Russia and China.[14]

Our study looks at these actions through an issue-centric lens to find that it is certain territorial concerns that drive Russian coercion. International politics and energy concerns might explain Russian action, but the domestic context cannot be left out. For Kubicek and Ambrosio, domestic politics explain external action.[15] The work of Ziegler and Berryman covers Russia's uneasy relationship with the West on the issues of energy and post-Soviet space in general.[16] Ziegler looks at the trust levels the EU has with Russia as a reliable supplier of energy and finds that there is much more to be desired in this shaky relationship. This has led to the EU looking to diversify its natural gas supply list as well as its transit routes that bypass Russia, which runs contrary to Russia's goals and national interest in becoming a resurgent power.[17] Berryman goes further and argues that part of the reason for the 2008 Russo-Georgian five-day armed conflict is that Russia wished to paint Georgia in a negative light when it came to Georgia being an alternative pipeline transit country that would supply Europe with oil and natural gas.[18]

Russian identity is important to the state, and Tsygankov asserts that the peculiarities of Russian national identity is the key factor in explaining why Russians mistrust the West and seek to reassert Russian power.[19] Tsygankov also uses the soft power concept and argues that Russian culture and language is now so ingrained in the former USSR that these countries will always be pulled into Moscow's sphere of influence.[20] Putin was a key factor in this process, and as his administration revived, for many, so did the idea of a great and powerful Russia able to protect and project its interests across the globe.

Other works on identity are important for our examination. Feklyunina's work looks at the concept of image, both self and other, for explanations of Russian foreign policy behavior.[21] The territories of the Tsarist Empire and then later Soviet Union help make Russia what it is, and the loss of control over this huge landmass and population since 1991 has given Russia somewhat of an inferiority complex. For Russia to be satisfied with what it is and how it fits into the world and its status as a world power, Russia must win back the hearts and minds of its former subjects.

For Feklyunina, Russia must use the power it has left, mainly through economic and cultural influence and dominance.[22]

Hopf looks at Russia's great power identity and how it is the legitimizing factor pertaining to the use of military action for foreign policy decision makers.[23] However, this process is complicated due to the notion that Russia has multiple identities that are produced domestically, between the state and society, and internationally. Russia under Putin has produced a Russia that is willing to use power politics to project his great power identity. As Hopf looks more at the concept of self when explaining Russia's foreign policy actions, Neumann uncovers the "other" concept of identity scholarship to explain Russia's actions.[24] He finds that as post-Soviet Russia has rejected the Western international norms of democracy and neoliberal economics, Russia has been rejected by Western states as a legitimate great power. Therefore, Russia and the West will conflict on most international issues, especially when it comes to issues over post-Soviet space. We use the concept of identity throughout this volume as a partial explanation of post-Soviet Russia's foreign policy behavior.

Russia has unique and different relations with each state of the former Soviet Union. Therefore, examining the bilateral relations between Russia and each former Soviet country may be important in explaining Russian foreign policy behavior. Perhaps the most complete volume that tackles the issues of Russia and the Commonwealth of Independent States (CIS), making up the post-Soviet space of the Near Abroad, is Bertil Nygren's work that looks at Russian foreign policy during the Putin era through the eyes of Putin.[25] Nygren splits the post-Soviet space into three basic security, economic, and cultural sub-complexes: the European (Moldova, Ukraine, Belarus), the Caucasian (Georgia, Armenia, Azerbaijan), and Central Asian (Kazakhstan, Uzbekistan, Kyrgyzstan, Tajikistan, and Turkmenistan). He then gives an overview of the bilateral relations between Russia and each state and discusses the outside competitors (the United States, NATO, EU, and China) and Russia's relations with these powers when it comes to political, economic, or cultural influence on these states and regions. Nygren's work is comprehensive and complete; however, his focus on analyzing Russian foreign policy in the CIS through the Putin prism reduces the generalizability of the project. It is our goal to consider the perspective not only of Putin but of other important actors and entities that influence Russian foreign policy as well as the point of view from the West.

In the end, it is likely that all these factors, plus others, combine to produce an outcome. Russian resurgent power is motivated by commodities prices, domestic political concerns, national identity, and also the need to keep the United States out of the Russian sphere of influence.[26] Post-Soviet space must be kept in Russian hands, and the 2008 conflict against Georgia was mainly motivated by the desire to keep Georgia from moving towards the West through NATO or the EU, rather than a desire to conquer territory. In some ways, the conflict was fought to teach Georgia a lesson and decrease its commitment to Western institutions. As we detail in this book, the opposite happened, and Georgia, although weary and bloodied, still tilts towards the West.

While these views are important and add key insights, they generally fail to achieve any sort of generalizable theory explaining Russian action or expanding how one might take lessons from Russia and apply them elsewhere. Our goal here is to not just explain Russian foreign policy but to also explain how all states with moderate power projection capabilities operate in the international system. In many ways, this book is not so much an examination of Russia but a social science application of our theory to the Russian region.

In addition to being generalizable, the theory here also builds on other empirical findings regarding the conflict processes of states in the modern international system. There is great interplay between the foreign policy analysis wing of the political science field and the conflict studies branch of the field. The conflict studies wing tends to be more broad, quantitative, and generalizable, while the foreign policy section tends to be more exhaustive in examining units as opposed to systems. These two modes of analysis can be blended together to build upon each other. The linkage between domestic politics and international outputs is clear but rarely put into practice.

This study falls more in line with Leng's thinking about foreign policy choices in their historical context.[27] We agree that coercive diplomacy more often than not fails, but we also point out that foreign policy alone is not made due to beliefs or expectations but also under certain international situations that we will elaborate in this chapter.[28] We also hope to avoid the trap of analyzing specific decisions in favor of examining objectives, goals, and entire grand strategic plans that a state marshals to deal with a foreign policy issue.[29] Our main dependent variable is not decision-making processes but general foreign policy outcomes as orientated towards particular states. It is in this sense that foreign policy is a dyadic and interactive process that cannot truly be examined through the analysis of one state alone.[30] Next we move on to an explanation of what tactics states choose and under what situations these contexts are in operation.

Russian Power Politics Tactics

Power Politics Tactics and Foreign Policy

Power politics responses are often thought of as the normal outgrowth of the national interests of states. When states interact in the system, they interact based on power politics considerations.[31] Power politics are a condition of foreign policy practices where force and power are utilized to settle disputes and exert influence on other states or actors. Vasquez defines power politics practices as "actions based on an image of the world as insecure and anarchic which leads to distrust, struggles for power, interest taking precedence over norms and rules, the use of Machiavellian stratagems, coercion, attempts to balance power, reliance on self-help, and the use of force and war as the ultima ratio."[32]

This view of foreign policy as part of the power politics practice is an unnecessary outgrowth of the perspective of international anarchy—an assumption that there is no assurance of security in an insecure world without a Leviathan. In that case, a state's best interest is served by building up its military and gathering allies when there are pressing security threats.[33] To deal with a threat, states utilize coercive diplomacy to deter the other side from acting or to compel them to change goals to match the pressing state's strategic interests. Power is thought to allow a state to do what it wants, take what it wants, and to flaunt the conventions of international norms.

The problem with this viewpoint is that it is theoretically unclear and provides disastrous policy advice at the foreign policy decision-making level. If these tactics are successful, more often than not, states learn that coercive diplomacy is the best way to achieve foreign policy ends while other states view this behavior as problematic and will eventually mount a counteroffensive to deal with the problem state. As Leng notes, if a hardliner coercive strategy works in one crisis, it then encourages the challenger to escalate tensions and refuse to back down in the next few crises.[34] Coercive diplomacy can work for a few occasions, but making this form of diplomacy the natural outgrowth of foreign policy decisions only leads to more conflict, rivalry, and strong enemies.[35] The outcome typically is the opposite of what was intended.

It is often said that Russia's only true friends are its army and navy.[36] This perspective is an outgrowth of a power politics mentality. In this way, power politics becomes a cultural point rather than a strategy. Power politics behavior becomes a way of life, a value and belief, or a perspective rather than an action taken to settle a problem. This form of strategic culture is problematic because foreign policy should be nuanced; it is the external expression of internal goals and should be a thorough reflection of the optimal strategy to deal with an international situation.[37] Instead, when power politics strategies become part of the strategic culture, these coercive strategies become the normal reaction to situations, a standard operating procedure under the context of a rivalry or a contentious territorial dispute. The actor is then constrained in its actions and choices.

Russia's foreign policy behavior is a power politics form of strategy and should not be taken as the optimal path. There are better paths and better courses of action. This work was not built to prove classical realism correct; instead, our perspective is that this paradigm is an outdated reflection of the Cold War that does not describe how successful foreign policies in the modern system operate. The security interests of both the United States and Russia are not served by relying on past tropes that served these states relatively well during the turbulent Cold War. Instead, a better strategy must be devised, a tit-for-tat strategy that deals with the situation as it manifests and considers the interests of the parties involved.[38] The goal should be to find where commonality exists or where foreign policy goals converge.[39] Threats, uses of force, and demands are not the best way to achieve foreign policy ends. States learn that entities that utilize these strategies will never be satisfied or appeased. Instead of trying to work with such states, a state will observe the power politics path in operation and defect to the other side (if there is another side) that offers a more reasonable perspective.

New Power Politics Tactics

Here we highlight how Russia uses new methods and tactics in light of its reduced capabilities after the Cold War. The focus is on cyber power, energy power, and the use of maritime power to coerce its enemies and friends alike to follow its demands. These forms of power differ in their application, but we argue that the outcomes remain the same. Where power is used to coerce, the aggressor state often fails to achieve its ends.

The first realm we analyze is cyber power. Russia is one of the most cyber-capable states in the world. It has the ability to marshal cyber power that few states have given. Examples of offensive actions are its cyber tactics used against Estonia and Georgia. The state has a sizeable population of tech-savvy citizens and military personnel ready to use cyber power for the state. Given that Russia is extremely nationalistic, it stands to reason that many will be willing to use this power for aggressive actions to promote Russian interests.

Another issue is that the FSB filters most international Internet traffic, meaning that all cyber activity coming into or out of Russian cyberspace goes through the organization's servers, suggesting they have complete control on how Russian cyber is used externally. If the use of cyber power might be detrimental for the state, the Russian bureaucracy will likely stop these actions and funnel interests towards activities that will benefit the state.

Given the contours of Russian cyber power, the question we have is what happens when these forms of power are directed externally to achieve Russian ends? We would then expect that Russia would often use its cyber power externally since attribution is less assured in cyberspace. Yet we often find the opposite. Russia and others are restrained in their use of cyber power.[40]

Next we examine Russian energy power. The huge Russian landmass has proven reserves of hydrocarbons, namely oil and natural gas. Countries that were once part of the Soviet Union were able to take advantage of these endowments, which, for the most part, came from the Russian SSR. Now these countries are independent and must pay for any natural resources coming from Russia. Those that do have their own natural resource supplies will usually need to transport their products through Russian territory and Russian-owned pipelines. Russia has used these new situations to advantage, usually in a coercive manner; and these foreign policy moves are what we call coercive energy policy.

Some post-Soviet states, such as Ukraine and Belarus, serve as transit territories to get Russian oil and natural gas to Russia's most important energy customer, the European Union. This gives these states leverage over Moscow, in that they can ask for subsidized pricing in return for cheap transit rates across their territories. Disagreements have been the norm over the years, and now Russia has built and is planning to build pipelines that circumvent these territories, so that Russian oil and gas can get to EU customers directly and with less political entanglements. Russian coercive energy policy, therefore, comes in many forms.

Here we argue that Russia uses its energy leverage over states of the former Soviet Union to bring these states closer to its political sphere of influence and away from other political orbits, especially the West. It has a variety of tools at its disposal: natural gas and pipeline transit pricing, the threat of circumventing natural gas pipelines around post-Soviet space, and coercing states with buying up pipeline infrastructures of these states in return for the forgiveness of some debt owed to Russia. Energy power is perhaps Russia's most important coercive tool in its arsenal of new power, and we devote two chapters (Chapters 5 and

6) to this topic in this volume. What happens when Russia uses energy power?

Finally, we look at how Russia uses its revitalized maritime power to achieve its ends in the post-9/11 world. This is a key question given that global warming is changing the dynamics of naval operations. Russia, therefore, has been aggressive in the navel domain in order to gain more resources and opportunities in the North. What happens when Russia takes action in this arena? Does it use power in an offensive manner, and do the means justify the ends?

The Coercive Path versus the Accommodationist Path

For a major power, a state with power projection capabilities and global interests, there are typically two choices for foreign policy. These choices represent two extremes. One path is the power politics–coercive diplomacy path and the other offering is the accommodationist–world order path.[41] While these perspectives are polarizing in the sense that they offer the extreme opposite positions, they also tend to reflect the reality of the options on the table for major power foreign policy decision makers. The middle path is often pushed as a path of inaction; instead, the debate tends to be between those who want extreme action and those who want to find a viable path towards cooperation.

The coercive diplomacy path, also called the hardliner path, represents the path where the use of force and escalating demands are the primary tactics of foreign policy discourse.[42] Demands are presented in a way that failure to comply will be met with punishment.[43] When there is a difference of opinion, the state that uses coercive diplomacy will respond with threats and escalation and seek external support against the challenger. Any deviation from the path can be criticized as not sufficiently accepting of the realities of power; the goal should be to compel the other side to alter its behavior. Challenges must be met with an extreme response to bend the other side's will. The assumption is that the opposing side will be rational and capitulate in the face of the determined power, but in the context of rivalry, salient issues, and a demanding public, capitulation is often not an option.[44]

States that rely on coercive diplomacy directed at challengers will typically not bend towards a compromising outcome when given the option in a negotiated setting. The outcome of choice is typically to maximize the positive gains for only one side. This path seems to follow Morgenthau's dictates. However, it also forgets the advice he suggested in that making foreign policy choices relies on understanding both the power State A has and the power State B has, but also what objectives

State A and State B have and where there might be a reachable agreement.[45] Taking a hardliner stance towards negotiations only ensures that one side will win or both sides will lose. Even if one side wins, the dynamics of conflict tend to ensure that there will be negative and devastating checks on the winner sooner or later.

The accommodationist path (also called the reciprocating strategy by Leng) has provided an able expression of foreign policy objectives throughout history.[46] It is not meant to suggest a path of capitulation but a path that seeks compromise above conflict. From this perspective, conflict is unnecessary and counterproductive to the goals of a state's foreign policy objectives and should be the option of last resort. Many seem to remember the Cuban Missile Crisis as an example of the efficacy of power and strong will where John F. Kennedy stared down the Soviets and made them give in. The reality of the situation was that the United States agreed to remove missiles from Turkey in exchange for the Soviet's removing missiles from Cuba.[47] Conflict in this setting would have possibly led to all-out nuclear war, obviously an outcome unacceptable to both sides; so compromise became the objective and the solution to the problem at the same time.

An accommodationist path would take each foreign policy situation in its context and make choices for a strategy according to the situation, generally finding the path of compromise the optimal solution to the conflict. The winning strategy would often be an agreement that suits both sides since zero-sum choices are rare in reality. Goal maximization can typically be found in a choice that sacrifices some minor gains in favor of general victory. Trust and partnerships are not built on relationships that take advantage of the opposing side. Even thinking of the other side as an adversary is a distraction best left to the world of sports, not politics. Through accommodationist foreign policy, rivalries can be avoided, and stability can be a model outcome for all states.

The choice between accommodationist paths and hardliner paths is a critical choice in foreign policy decision making. Conflict is hard to avoid if both sides are hardliners and utilize coercive diplomacy.[48] The question then is to figure out why this situation of dual hardliner stances develops, and how to avoid its entrenchment in the system. To achieve these ends, we must first understand current theoretical sources for foreign policy action and then develop a new system of evaluation that takes the context and the issues at stake as the central part of a foreign policy portfolio. By understanding the context of the inputs for foreign policy choices, we can then move towards understanding and evaluating the foreign policy of specific states.

Realist Theories versus Constructivist Interpretations of Foreign Policy

As Rose notes, "theories of foreign policy seek to explain what states try to achieve in the external realm and when they try to achieve it. Theory development at this level, however, has received comparatively little attention."[49] It is suggested that this is due to the popularity of the structural focus of Waltz's neorealist theory.[50] However, structural neorealism cannot explain the behavior of individual units in the system, a fact Waltz himself admits. It predicts stable outcomes: either a state attempts to balance power or the state is crushed, and this leaves little room to examine and study the motivations for state action in foreign policy.

Attempts to cover this deficiency have resulted in few solid advances in the field. Offensive realism predicts that goals of power maximization will dominate the system, but this prediction fails a simple recent validity test through an examination of such countries as China or the United States since the end of the Cold War.[51] It also fails when compared to international state history.[52] Defensive realism focuses on the more balanced deterrent aspects of power projection abilities, suggesting that focusing on the defense rather than the offensive is the path to international stability.[53] While defensive realism has promise in the realm of strategy, it does not help us explain foreign policy action over contentious issues.

A relatively new path called neoclassical realism has emerged, but it suffers from many of the same problems as its predecessors. It is an attempt to recover the wisdom of classical realism advanced by such thinkers as Morgenthau, Machiavelli, and E. H. Carr.[54] Neoclassical realism focuses on perceptions of power by elites and decision makers rather than pure considerations of material power in determining foreign policy action.[55] This theory is an advance on others in that it does not take systemic or material factors as the primary focus of foreign policy objectives. Unfortunately, it fails in that it does not outline how exactly perceptions are built and which factors feed into views of power considerations. Anything and everything may count in the construction of oppositional viewpoints in foreign policy conflicts, according to neoclassical realists. This reality provides no firm basis for prediction or theory, only a general method of storytelling in explaining past processes. As Rose himself warns, "Critics might see the school's emphasis on perceptions as a giant fudge factor, useful for explaining away instances where foreign policy and material power realities diverge."[56] Perceptions and other socially constructed factors are used to cover the gaps in the theory; yet these factors should be built into the theory in the first place and be the basis of prediction rather than an afterthought.

The goal should be to build a theory that explains the past, present, and future of foreign policy action.[57] Theory must be empirically accurate, but it also must speak to policy debates and future paths. The neoclassical realist theory also fails to understand the concerns of prudential realists like Morgenthau and Machiavelli, who advocate a more restrained form of foreign policy.[58] Power should be marshaled for the defense of the homeland; when states go beyond this basic idea, they may unleash more than they are prepared to confront.

All of these theories fail to explain how states normally operate. They tend to take assumptions of insecurity beyond their natural boundaries in the system as it is currently constructed. Conquest is dead, and conflict in the system is declining.[59] Instead of building a system of foreign policy interactions based on pure power considerations, it would be more beneficial to examine the construct of foreign policy in the modern system according to the contexts of the state in question, the ideas and values of the state, and the past history of conflict the state has experienced. Together, these elements can help determine what foreign policy objectives a state will have. A basic notion of power projection to encourage state behavioral modifications either through threats or a strong defensive posture does not link up with how states build their foreign policy objectives. No one would argue that Israel builds its foreign policy objectives based on the military power it holds alone, and its specific capabilities far outweigh other states in the region; yet Israel operates under some considerable fear from small powers and minor terrorist threats because of the potential damage these relatively less powerful actors may cause either through direct attacks or insurgent terrorist violence. Power alone does not explain strategy, nor does it explain foreign policy choices. It is unfair to place this burden on realism since the early practitioners (like Waltz) warned against this, yet this is the direction in which modern scholars have pushed the theory.[60]

A better path is needed. This path must account for the context in which the state operates and with how these contexts affect the behavior of individuals in the unit of interest. How a state perceives threats is not enough information on which to base a viable theory of foreign policy. This is where constructivism is a useful lens.[61] Predictions need to be more concrete to deal with the setting of the security situation as it is constructed. There can be a perceptive theory of foreign policy action that accounts for power under certain contexts. Instead of the defensive realist focus on deterrent processes, most states in the system feel no need to build defenses against a nonexistent threat. Instead we have measures of action that can be used to construct a measured foreign policy pathway according to the situation states will find themselves in and the history of past conflicts. This is the outlook we use for this volume.

The choice between accommodationist and coercive strategies must be made under the contexts of rivalry and the issue in question. Foreign policy strategies are made through how a specific reality is constructed.[62] Scholars should go back to the early ideas of Snyder and Rosenau.[63] Perceptions do matter, but how the setting is constructed is critical to explaining which options are on the table to deal with foreign policy situations. Much more than power needs to be considered in the process of evaluating foreign policy choices, and the next section endeavors to develop such a theory.

The Sources of Coercive Diplomacy

Situational Coercive Diplomacy

Coercive diplomacy is not the natural state of international politics. As Jack Levy defines the term, "it is a strategy that combines threats of force, and if necessary, the limited and selective use of force in discrete and controlled increments."[64] Some would advance the idea that whenever states interact, they interact through the lens of coercive power politics tactics.[65] Clearly this is not true; when Costa Rica interacts with Brazil, it does not do so on the basis of power politics considerations and the use of force but instead based on the goal of normal communication and typically benign cooperation. Their interactions are based on the hope of advancing the agendas of both countries. Militarism rarely comes into the equation for most foreign policy interactions; it is generally the domain of last resort or for specific situations. Coercive diplomacy, the threat to use force, and hardliner stances are used by a particular subset of states that operate only under certain contexts.

Power politics tactics tend to manifest in the nexus of two contexts: rivalries and salient issues. Power politics tactics both lead to and sustain an international rivalry.[66] Under the context of a rivalry, power politics responses and coercive diplomacy become the normal practice of interaction. The background of the conflicts between two states needs to be examined to understand the current path and challenges states face. The past shapes the future. Conflict events, even in the distant past, influence how states respond to contemporary challenges. Rivalry is defined as a situation of animosity where the issue positions of the contenders are orientated towards the other in order to deny a gain to

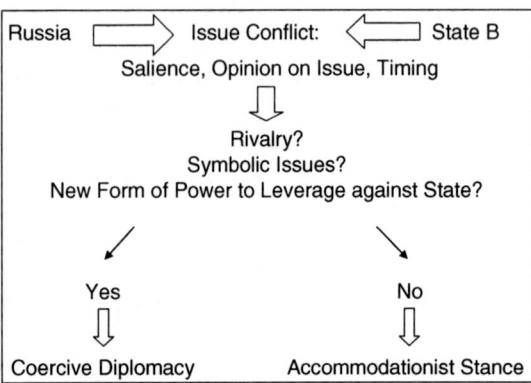

Figure 2.1 Sources of foreign policy outlook

the enemy. Empirically, it is measured either through historical perceptions or through the number of crises and disputes a pair of states have engaged.[67] Certain factors help produce a rivalry outcome. These factors are alliances, arms races, territorial disputes, dissatisfaction, and newly independent states.[68]

The other form of contextual interaction that becomes important is salient critical issues that take on symbolic stakes. These issues are typically territorial issues, but they can also take the form of transcendent issues, such as regime leadership, hierarchical power relationships, or political ideology in a region. Here we will explain the contexts critical to the making of foreign policy outputs so as to better understand when states choose the power politics path. Figure 2.1 summarizes the path constructed here.

Issue at Stake

The potential issues at stake between states are a critical element of foreign policy that often is left out of the equation in other theories. States only engage in foreign policy areas that are connected to their portfolio of interests. While states will have a myriad of interests and concerns in the system, most focus on a few issues at once. Which issues matter to a state is determined by the types of issue on the agenda, the salience of various issues, the time period in which the issue is raised, and the public and elite reactions to these issues.

A basic typology of issues would be focused on territorial issues, ethnic concerns, energy distribution and politics, human rights concerns, general policy objectives, regime leadership issues, and environmental issues. These seven issue areas cover the majority of the differences of opinion between states in the modern international system. Territorial issues are the most problematic and conflict prone.[69] The reason for this is the symbolic nature of territory. Often territorial disputes invoke the homeland or a life-death struggle. Compromise is not an option when the stakes are so high.

Symbolic and transcendent issues are the most conflict-prone issues in the system.[70] The logic behind this is that concrete stakes can be easily solved and divided with rational bargaining strategies. Symbolic issues defy bargaining since the stakes are intangibly high for each side. The issue becomes a question of the status of the state and its place in the system rather than a position in an argument. Losing means a loss of face and a demonstration that the state cannot protect its interests and security. These types of issues transcend simple cost-benefit analyses.

This form of issue construction is prevalent in any issue concerning post-Soviet space. For many in Russia, the status of the region is purely a Russian question and not an international question. This region is the special providence of Russia, and any incursion by other powers represents a failure on Russia's part to exert its power projection abilities. While not a traditional territorial issue in the form of a boundary question, the question of mastery of the post-Soviet space is an issue that is territorial in nature because the region holds historical significance for Russia. It took Tsarist Russia centuries to secure a "Near Abroad," and contemporary Russian leaders are loath to be associated with its diminishment or loss. Russia feels that it has legitimate domain over this region and other states have no legitimate standing in the region. In this form, the conflict becomes an "us versus them" competition for control over a region.

Which issues become transcendent and important in the foreign policy discourse is usually determined through the interplay of two related questions: What is the salience of the issue? and In which direction do elite and public sentiments flow? Salience is simply the strength of interest in relation to other questions on a certain issue. A highly salient issue will be important to the public, elites, and to the status of the state as an entity.[71] Failure to protect the outcome of issues of high salience signals greater failings in the process as well as the status of the leadership of a state. There is a strong linkage between foreign policy goals and domestic concerns. Without the protection of the sacred core interests of a state, usually its homeland and general national ethos, it is questionable if the state really is the master of its domain.

Salience interacts greatly with public opinion, but it is not dominated by the factor. Issues can only be salient with the support of the public, but not all issues deemed important by the public are salient, and public moods can be directed by elites. The public can push issues towards the general agenda of a state, but sometimes the public pushes issues that are not strategically important to the state. It is here that the type of state in question becomes important. Authoritarian states have greater latitude to ignore issues deemed important by the public, while democratic states are more beholden to the whims of public sentiment. As it stands, no matter which state is being examined, the foreign policy processes they exhibit are not dominated by a unitary rational actor but by other actors who may be operating under tense, ongoing conflict situations.

Morgenthau's early thoughts on issues remain important today.[72] Issues important to the public can be added to the agenda by elites. They can control the destiny and preferences of the public through the use or control of information. This is why it is important to survey the sentiment of elites in addition to the public at large. The interplay between the two can be in the form of following the will of the public or determining the path of public demands, depending on the skill of the elites trying to construct the state's foreign policy portfolio.

History cannot be left out of the equation. The past history of interactions is an important variable in determining which issues are part of the state's agenda. Some issues have been dealt with in the past and are shifted towards the bottom of the interests important to the state because they are either nuisance issues or are just grandstanding attempts by state leadership. Other issues come up so frequently in certain contexts that any trigger under that context can lead to escalation with another state. The idea of rivalry in foreign policy codifies the importance of history in determining the issue portfolio of states.

Rivalry

A rivalry is a long-term animosity between a set of actors with some expectation of conflict. Diehl and Goertz operationalize the term by looking for repeated conflicts with a certain degree of competitiveness and a connection of issues at stake.[73] Historically, rivals are the most conflict-prone pairs of states in the international system. When rivals come into conflict, the entire region will shudder and small fights become wider conflagrations.

An important concept illustrative to rivalry is addiction. Rivals are locked in to conflict in that conflictual relations become the normal standard of behavior within a rivalry dyad. The relative positioning of each actor is also a key factor in rival dyads. All proposals are based on a zero-sum calculation where either side will attempt to prevent any public or private goods from falling into the hands of the rival. Anything and everything can be fertile grounds for conflict among rivals.

Thompson provides the most extensive typology of rivalry to date by identifying three main types of rivals.[74] They can be spatial (rivals disputing territory); positional (rivals contesting shares of influence in a system or subsystem); and ideological (rivals contesting the relative virtues of different belief systems). Huth, Bennett, and Gelpi also develop a typological system of rivalries that can be utilized for an examination focused on great power rivals.[75] They state: "the basis of a rivalry may be competing claims to sovereignty over territory or influence over the economic, political, or military policies of another state."[76] Therefore, we can classify rivals in this analysis as territorial or policy. Policy rivals can be further subdivided as either regional or geopolitical rivals.

During its duration, a rivalry can shift in type. Although rivals can span many issues, this analysis takes note of the *dominant* form of relations, as does other recent research on issue concerns. Most importantly, a rivalry does not require a connection of issues or types over its duration but a continued perception of threat.[77] Through the long lifespan of a rivalry, situations change and the issues under contention shift, but due to the dyad's addiction to conflict, the rivalry is difficult to resolve.

Shifting from a geopolitical policy rivalry to a regional policy rivalry allows for the perpetuation of conflict between rivals, at a different global context, but the rivals remain in a period of contention even after drastic changes in policy. For a rivalry to be in operation, some form of persistence or addiction must be evident and notable in relations. Relative positioning should dominate strategic concerns. Finally, does the conflict reflect positional (policy) or spatial (territorial) elements in its processes?

Not every state has the ability or the willingness to secure its geopolitical preferences. Russia is one such state in that its ability to affect action tends to be regionally focused on the former Soviet Union. Even the Arctic, where Russia has intense interests, is a regional issue in that Russia has the most Arctic seaports and the most underwater territory in the region (see Chapter 7). Therefore, the region under consideration is an important factor for the rivalry question. Location dynamics matter in that regions far from the homeland lack the salience to advance a rivalry situation, unless the state has hegemonic ambitions. At least as of the current period, Russia does not have hegemonic ambitions, despite what pundits and campaigns might argue.

In terms of foreign policy objectives, rivals provide the context for the majority of conflicts in the international system. Devoid of context, disputes can be settled without having to account for the past history of interactions in solving the problem at hand. States are free to build foreign policy portfolios unencumbered by past mistakes and challenges. In the context of a rivalry situation, any dispute between rival states has a high chance of escalation. All sorts of affronts to a rival are seen as a grand challenge to the status of a state. Challenges in the context of a rivalry must be met with equal or escalating force or the leader of the state will be seen as weak and capitulating towards his or her enemies. Coercive diplomacy is not a tool of all states; it is generally the tool of states engaged in rivalries that have contentious salient issues at stake.[78]

Foreign Policy Dynamics and Arenas

The Dynamics of Regional Powers

Russia is now a muted power. It will not reach the global apex of the power it held during the Cold War anytime soon. Instead, Russia has shifted its goals to be a regional power rather than a global power. It still has global ambitions, but its ability to be relevant in the far-flung reaches of the world is nonexistent in most places. Where Russia still is relevant globally is generally tied to its past relationships rather than current abilities. Russia is important in such states as Cuba and Syria, but its relevance is tied to the sale of weapons systems that other powers will not export, or past Cold War ties.

As noted in the prior section, the choice between hardliner and accommodationist stances is conditional upon the issue at stake and the past history of conflict. To evaluate and examine Russian foreign policy interactions in the post-Cold War world, we have chosen to examine all of the relevant arenas of foreign policy challenges that Russia faces. We exclude ethnic tensions and internal conflicts that have more to do with internal politics and Russia's domestic historical processes than international interactions. Here we will briefly explain why each arena is important and what we hope to gain from a thorough examination of the arenas in the context of Russian foreign policy.

Arena 1: Rivalry

The first critical question involves just what is a rivalry and how would one know when a rivalry is terminated? Through an analysis of Russia and the United States (Chapter 3), we are able to both explain how rivalries work and also determine the current path of Russian foreign policy. Many argue that the rivalry between the United States and Russia terminated with the end of the Cold War; we argue otherwise. The goal here is to understand the state of relations between Russia and other various states in the region. The rivalry of war and conflict is the most critical domain of analysis since the consequences of war far outpace the potential impact of other domains of interaction.

Arena 2: Public Opinion

To better understand the context of public demands according to various issues of interest, we conduct an extensive review of past public opinion polls that gather the sentiments of Russia and other regional actors to uncover the dynamics of opinion in Russia (Chapters 3, 6, and 7). Hutchinson and Gibler suggest the salience of territorial issues for various publics around the world is greater overall compared to salience for other international issues.[79] Gibler, Hutchinson, and Miller also find evidence for individual-level feelings of high salience over territorial issues.[80] The salience of various issues is determined, in part, by how the public feels about the issues. Here, we can pinpoint which issues control the agenda when Russia interacts with other powers. This analysis provides an important window and background to one of the critical contexts that determines foreign policy strategies. In some cases, elites set the foreign policy agenda; elites can direct public sentiment or utilize ongoing expressions of public will for their own ends. We also note that a rivalry cannot be considered terminated until the public accepts its termination. It matters very little if the leaders of both states proclaim the dispute dead as long as the public continues to perceive a rivalry and threat from the other side, which opposition parties can utilize to signal the current regime has lost touch with the security situation. Ending a rivalry is as much about changing the strategic culture of a state as it is about signing treaties and declarations of friendship.

Arena 3: Cyberspace

Some term cyberspace the new dimension of conflict, and therefore a new issue space. Russia has been one of the most active users of cyber power, and it is important to critically analyze how and when Russia uses the strategy in the realm of foreign policy (Chapter 4). Russia is less willing to use all aspects of its military and computational power even though other states seem to use the tactic more often and more liberally. While not a traditional territorial issue in the form of a boundary question, the question of mastery of the Near Abroad is an issue that is territorial in nature according to the idea of spheres of influence.[81] Yet Russia has restrained itself from using its capabilities in cyberspace in this region and beyond. Why?

In an analysis of Russia's actions towards Georgia, Estonia, and the United States, we can examine how a new modern power politics tactic is used and conducted in the system. Russia uses the tactic as a form of coercive diplomacy, but only against its rivals. Using the tactic outside of the rivalry situation makes little sense since the tactic of cyber attacks is so dangerous and path breaking that it could open Pandora's box in terms of cyber blowback towards Russia. Instead, when Russia does choose to use the tactic, it does so at such a low level that one must wonder about the value the tactic has in the future and when or if a state is determined to use it to its full potential. Some examples would be helpful; everything seems vague in the discussion of cyber stuff.

Arena 4: Energy Power

Perhaps the most evidence of use of coercive diplomacy is found in the arena of energy politics (Chapters 5 and 6). In order to achieve its objectives and utilize the power that is available, Russia has been using coercive tactics in the energy arena for some time. It is the pioneer of the use of resources as a tool to achieve its foreign policy ends.

Through a study of this arena, we can evaluate how and when Russia uses energy power. We can also uncover if the tactic is successful in compelling the other side to bend to the will of Russia. As is the case with most uses of coercive diplomacy, sometimes the tactic is effective, but when used repeatedly, it tends to fail and results in states choosing a path of self-reliance and conflict rather than capitulation to aggressive demands.

Arena 5: The Arctic

With the impact of global warming, the region is now a contested space between Russia and other capable powers. The question we have relates to how Russia is using its power in this region (Chapter 7). This arena serves as an interesting test of our general theory in that there are few rivalries operating in the region, and the Arctic is not a salient foreign policy issue for most states, according to our salience score or public opinion measures; therefore, we would not expect Russia to use its full power politics might to achieve its ends in the region. In the absence of our contextual variables, just what sort of foreign policy does Russia utilize? How Russia interacts with other states in the region might be a harbinger for how Russia will interact in the system when issues of the former Soviet Union are not on the agenda.

Our arenas are comprehensive but not mutually exclusive. We cover all aspects of Russian foreign policy that have the potential for conflict and the use of force. We cover international law and institutions where these issues come up in various arenas but do not devote a specific focus to these ideas because it fits our analysis to handle these issues in their proper context. Norms and institutions are covered, but this is not a constitutive analysis.

While Russia is the focus of this examination, the theory presented here is intended to be generalizable to most middle powers, states like Nigeria, Brazil, Iran, and Japan. All of these states have significant security interests and abilities but lack the geopolitical engagement of hegemonic states. These powers are vying for regional dominance and usually are looking no further. China's place in this theory is debatable, as it currently falls in the middle power path but is quickly moving towards having a geopolitical orientation. The key condition for our theory is a regional focus rather than a global outlook, and Russia fits this qualification.

Conclusion and Path Forward

As Russia attempts to reemerge, it faces serious challenges to its ability to influence and control its regional sphere of influence. Russia's agenda is no longer global but regional. Its ability to project global power has declined; instead, Russia must focus on how it interacts with former Soviet states and past Cold War enemies over various issues that are critical to its ability to remain influential in the system. Russia now interacts with the world using new sources of power, including cyber and energy power.

The basics of the framework advocated in this chapter are straightforward. The choice between the use of coercive diplomacy and accommodationist strategies occurs under the contextual settings of the critical questions on the foreign policy agenda. The main elements of context in the system are the issues under contention and the rivalry settings critical for actors with power projection capabilities. By understanding the power and tactics available to Russia and the setting in which it operates, we can understand how choices about the future direction of policy are made.

The rest of this book will be focused on evaluating how Russia uses its foreign policy abilities and why. The situation under which Russia operates is important for answering the most basic questions critical to foreign policy debates. Will wars and conflict occur? Who are Russia's rivals, and how does it interact with them? Which new tactics does Russia use and under which arena? These are all questions that can be evaluated based on the framework established here and hopefully can be extended by others to different regions and actors throughout the world.

Power politics strategies are a choice dependent on context rather than whim or compulsion. Understanding this choice is critical in understanding and evaluating foreign policy objectives and outcomes.[82] Without providing some sort of theoretical motivation for action, analysis of the ends and means of Russian foreign policy is empty in that the analyst is relying on hunches and speculation rather than a well-grounded theory. Chapter 3 looks in more detail at the dynamics of the "new" US–Russian rivalry, which is different in scope and magnitude than the Cold War-era superpower rivalry.

3
Rivalry Persistence and the Case of the United States and Russia: From Global Rivalry to Regional Conflict

Introduction

Researchers have noted that a certain group of states account for a disproportionate amount of conflict in the international system.[1] These states are rivals, or long-standing enemies, addicted to reducing gains for the other side and apt to challenge the goals of their rival. We now know that rivals are highly dispute prone, experience frequent territorial disputes, and can disrupt international and regional power systems. Despite progress, current theories regarding the termination of rivalries appear incomplete because they ignore domestic factors and the "principal" rivalry concept. Our central premise in this chapter argues that theories of rivalry termination are underdeveloped and fail to account for the settlement of outstanding issues and sources of rivalry persistence at the domestic level of analysis. If both the issues at stake in a rivalry remain unsettled and perceptions continue to be adversarial, the rivalry situation will persist. If the rivalry situation persists, an important context that dictates foreign policy action is in operation. This provides the context for Russia's use of new foreign policy tactics. Our argument is that the rivalry between Russia and US continues but has shifted in scope and fits with the reduced reliance on conventional forms of power.

To argue our point, we suggest that the rivalry between the United States and Russia shifted from a geostrategic rivalry to a regional rivalry, defying termination predictions. This case is important because the persistent antagonism between the US and Russia will remain a source of turbulence in the post-USSR region and system for years to come. This conflict provides the context important to understand the foreign policy choices of states.

The US-Russian rivalry demonstrates the enduring nature of historic animosity and that misperceptions rooted in history can cause a rivalry to endure even if one state seeks to change the relationship. Rivalry imposes limitations on conflict resolution and the construction of stable institutions in regions of dispute, reminding researchers of the importance of this topic.

This chapter also looks at elite and public opinion to examine the process of rivalry termination and specifically, the evolution of the US-Russian rivalry.[2] This rivalry is an important test case since it provides the motivation for many theories of rivalry dynamics, particularly termination theories in the post-Cold War era. We suggest that the rivalry was not terminated with the end of the Cold War and the subsequent post-communist reforms in Russia; these shocks did little to end decades of animosity, particularly on the Russian side. A process of evolutionary learning and historical misperception has created a relationship of potential conflict between the two states. The US-Russia rivalry continues to be the "principal" rival for both states and has only changed from a geopolitical rivalry to a regional policy rivalry with the potential of escalation after disagreements.[3] The conflict between Russia and Georgia, as well as the escalating divide between the US and Moscow over the civil strife in Ukraine, highlight the importance of this analysis and remind us that rivalry termination theories are important points of study for those interested in the relationship between military strategy and diplomacy on the one hand and society on the other.

While there have been efforts to quantitatively account for rivalry termination, the study of interstate rivalry needs to be advanced through the application of historical and contextual analysis of a rivalry's domestic political situation in order to understand termination dynamics prior to statistical investigations. Using the issue-based approach, this chapter does just this. In sum, the question we ask is twofold: 1) Did the US-Russian rivalry terminate circa the 1989–93 period? and 2) What continues to be the impact of societal and regional forces on the status of the rivalry?

The Concept of Rivalry

A rivalry is long-term animosity between a set of actors with some expectation of conflict. A rivalry can simply be classified as a pair of states who engage in several militarized interstate disputes (MIDs) within a certain timeframe.[4] An important concept illustrative to rivalry is addiction. Rivals are addicted to conflict in that conflictual relations become the normal standard of behavior within a rivalry dyad. The relative positioning of each actor is also a key factor in rival dyads. All proposals are based on zero-sum calculations where either side will attempt to prevent any public or private good from falling into the hands of the rival. There can be spatial (rivals disputing territory), positional (rivals contesting shares of influence in a system or subsystem), and ideological (rivals contesting the relative virtues of different belief systems) rivalries.[5]

A typological system of rivalries can be utilized for an examination on great power rivals.[6] "The basis of a rivalry may be competing claims to sovereignty over territory or influence over the economic, political, or military policies of another state."[7] Therefore, we can classify rivals as territorial or policy.[8] Policy rivals can be further subdivided as either regional or geopolitical rivals. During its duration, a rivalry can shift in type. Although rivals can span across many issues, this typology takes note of the *dominant* form of relations, as does other recent research on issue concerns.[9] Most importantly, a rivalry does not require a connection of issues or types over its duration but a continued perception of threat. Through the long life span of a rivalry, situations change and the issues under contention shift, but due to the dyad's addiction to conflict, the rivalries are difficult to resolve.

Shifting from a geopolitical policy rivalry to a regional policy rivalry allows for the perpetuation of conflict between rivals, at a different global context, but the dyad remains in a period of contention even after drastic changes to the policy. The task that remains is to apply concepts of rivalry to cases or data. For a rivalry to be in operation, some form of persistence or addiction must be evident and notable in relations. Relative positioning should dominate strategic concerns. Finally, does the conflict reflect positional (policy) or spatial (territorial) elements in its processes?

Rivalry Persistence and Termination

The analysis of rivalry as the dependent variable allows for the termination of the relationship to be the subject of interest. Goertz and Diehl believe rivalries are stable conflict patterns that require a "shock" or explosive change to terminate the relationship.[10] Their theories consider both exogenous shocks, such as world wars or major territorial changes, and endogenous shocks, such as civil wars or regime changes, to be the two main causes of rivalry termination. Additional research has demonstrated that the terms of settlement of the last dispute can have a large impact on the probability of rivalry continuation and maintenance.[11]

In opposition to Goertz and Diehl's requirement of a political shock for rivalry termination, McGinnis and Williams suggest that a superpower rivalry differs from other types of rivalry in that it demonstrates stability despite political or technical changes.[12] Huth et al. also find in their examination of great power rivalries that systemic theories do not trump analysis at the domestic level.[13] To understand the interactions of rivals, scholars should look beyond explosive exogenous and endogenous shocks and incorporate analysis of domestic variables to determine theories about rivalry duration.[14]

Bennett integrates the previously discussed theories of Goetz and Diehl and his earlier tests of rivalry termination theories into an all-inclusive model.[15] Bennett, along with Cioffi-Revilla, suggests that the probability that a rivalry will terminate increases with time.[16] Time allows rivals to attempt dispute settlement and allows generational attitude change within the political leadership. In general, Bennett finds that rivalries with high issue salience last longer, while regime change increases the probability of termination.

To determine the accuracy of current hypotheses of rivalry termination and evolution, we ask the following questions: Has the US-Russia rivalry terminated as the existing literature claims? Furthermore, how have domestic and regional concerns impacted the status of the rivalry?

The application of this case study will help stimulate solutions to general theoretical problems and contribute to our understanding of how rivalries terminate.[17] The US-Russia rivalry is an important case since it has not produced a direct war, which is a rare occurrence for enduring rivals. This case is also important since the end of the Cold War produced political shocks that were coupled with Russian regime change in 1991, events that current literature would suggest terminated the rivalry. Yet, these exogenous and endogenous shocks did not produce the expected outcome, and most scholars currently code it as enduring.[18]

We believe that current theories of rivalry termination are incomplete in explaining why some rivalries endure. Rivalries do not always end with political shocks, democratization, or war. Rivalries are ingrained in the foreign policy identity of nation-states through learning from experience, which sometimes is abetted by propaganda. As Goertz et al. put it, rivalries "lock-in" and are hard to dislodge.[19] Even if the political system of a state changes, the elites will still have been raised and conditioned in an atmosphere of rivalry. This often leads elites to resort to hostile relations when challenged by a historic enemy.

Research in this chapter suggests that interstate rivalries are a learned process that can only be unlearned or terminated in three ways: 1) issue resolution, 2) serious military defeat, and 3) rivalry linkages. The most common termination mechanism occurs through the resolution of the main sources of conflict between the two states. A rivalry will end if the leadership of both states agree, have public support, and resolve the main issue of contention in a mutually satisfactory way. While similar to Bennett's operationalization of rivalry termination, this rule does not require a formal treaty of settlement. A treaty settling the issues of a rivalry would be helpful in coding, but treaties are not a necessary condition, in that a verbal agreement will suffice if accepted by both domestic elites and the mass public—a key caveat. While settling the main issue under contention will likely settle a rivalry, this is not the case for all rivals. Some rivalry behaviors are so ingrained in the *modus operandi* of a state that rivalry might endure even if the main issue is settled.

The second way a rivalry might end is through serious military defeat. Examples of rivalry termination in this manner include Germany after World War II and the Ottoman Empire after World War I. The collapse of the state as an international military actor can preclude the extension of rivalry relationships. It was widely thought that Russia has ceased to exist as a great power actor after 1991. Russia's increased economic, energy, and military prowess throughout the former Soviet Union after 2000 demonstrates that it remains a great power and has only lost its superpower status.

The third and final mechanism for rivalry termination is through a process of rivalry linkages.[20] Using Thompson's concept of a principal rivalry, it is reasoned that a state can only have one principal rival at a time.[21] When a new challenger develops, this challenger might develop into a principal rival. This new rival will require a new orientation of foreign policy goals, including ending or reducing the levels of crisis with other rivals. A state can have a number of rivals at one time, but it can only afford one principal rivalry because it requires the state's full attention to deal with said enemy.

We argue the US-Russia rivalry did not terminate after the end of the Cold War because it has not met any of these three conditions for rivalry termination. There has not been a resolution of issues. There was an attempt to settle issues in dispute between the two states, but some elites and parts of the mass public (mainly in Russia) continue the rivalry over regional concerns. There has not been a direct war between the states, so the second point concerning serious military defeat has not occurred. Third, and finally, neither Russia nor the United States has developed new principal rivals.[22] Logically, the locus of rivalry continuation is on Russia because the changes thought to bring about the end of the rivalry mainly happened in Russia. When these changes did not have the desired effect, the rivalry endured.

Convergence of preferences between the public and elites may also be a necessary condition for the termination of rivalries. This point is covered in detail later in this chapter. Currently, Russia considers the US as a rival due to the latter's influence in the former Soviet states. Equally, the US considers Russia as a potential problem in the stabilization of nascent democracies. Washington's other concern is over regional stability as Moscow actively supports separatist enclaves in Georgia and Moldova. Russia viewed the democratic governments of the "Color Revolutions" in Ukraine and Georgia as direct threats to its territory and consequently, uses energy supplies as a form of power. For Ukraine, its turn towards the West and a more stable democracy has been met by Russia with destabilization efforts that include the annexation of Crimea and the indirect support of ethnic Russian separatists in Ukraine's eastern region that has brought the country to the brink of civil war.[23] The US and EU have responded with incremental economic sanctions on Russia's powerful individuals (asset freezes and travel bans) as well as its major industrial sectors.[24] Russia's coercive behavior in post-Soviet space since the fall of the USSR has made the US concerned with the extent of Russia's monopoly over energy supplies to Europe. The split over Kosovo is arguably most illustrative of the policy differences between these two states: Washington aligns with Pristina while Belgrade enjoys support from Moscow. Rogov even goes as far as to suggest that cooperation between the states cannot fully develop until they develop a mutual common enemy.[25] This process has continued over the status of Syria. Until the United States and Russia resolve all these issues and change their strategic culture of rivalry-focused foreign policy, the rivalry will continue. This leads to two hypotheses:

H1: The US-Russia rivalry continues past 1991.

H2: Concern for regional issues has led the rivalry to shift to a regional-policy rivalry.

The US-Russia rivalry will serve as a crucial case in that it provides a relevant historical and contemporary example that many theorists are familiar with and also generated many theories of rivalry relations in the first place.[26] We will now shift to an examination of the sources of rivalry persistence to investigate our hypotheses. An examination of the issues relating to post-Soviet space and international issues of contention will demonstrate that the rivalry endures to the present day.

The Evolution of the United States-Russia Rivalry

Depending on the starting point, the US-Russia rivalry has continued for about 60 years. The initial conflict was over the distribution of influence throughout Europe after the fall of Germany. The rivalry began as a regional policy rivalry and became a series of global policy disputes. The ascension of Gorbachev in 1985 signified a new phase in the relationship between the two states. By 1990, an agreement was reached over the reunification of Germany. The period's apogee of cooperation was in July of 1991, when the Strategic Arms Reduction Treaty (START) was signed. This treaty reduced the nuclear arsenals of both countries and committed each state to external measures of verification.[27] However, this period of cooperation did not last long. During the mid-1990s, the Russian-US relationship grew cold again as the two states began to compete and clash over regional rather than global issues. Now it is to the point where the United States will not ratify updated nuclear arms reduction treaties, and Russia has canceled the Luger-Nunn deal that helps discard nuclear materials.[28]

The settlement of outstanding issues and alignment of popular perceptions seem to be more pressing concerns for the United States and Russia. A close examination of the Russia-US relationship indicates that the rivalry has not been terminated. Formal attempts to end all disagreements have not resulted in a cooperative relationship required to terminate a rivalry. Official reforms were made from the years 1991 until 1993 but since 1993, the relationship has grown more conflictual and rivalrous. The sources of this development can be located in the realms of domestic politics and regional issues.

Cooperation and Post-Cold War Relations

The absence of war between Russia and the US throughout their rivalry shows that some form of cooperation existed. Midlarsky et al. write, "[interactions] gave rise to a set of beliefs as to what could or could not be done without risking war. Second, these tacit understandings, sometimes worked out through trial and error, led to a mutual realization immediately after the Cuban Missile Crisis that certain forms of behavior and competition were too risky and that efforts should be made to mitigate potentially dangerous interactions."[29] Levy's analysis of foreign policy learning shows that for lessons to be learned, they have to be experienced. In 1991, new leaders came into power that had not experienced Cold War cooperation firsthand.[30] While some "old hands" remained, the vast majority of those who came into leadership positions had never experienced such relations during the height of tensions (1960s). A policy change may have actually increased the chances for rivalry continuation because past experience and learning was nullified by new leaders. The lack of firsthand experience by elites in the new Russia, coupled with domestic politics, explains why the rivalry was not immediately terminated in 1991.

In the early post-Cold War period, Russia abandoned its commitments to an empire, gave up hopes of military parity with the US, and expressed desire in either joining NATO or creating a new institution like it.[31] The first step towards a new cooperation between the rivals required a reworking of the alliance structures that defined the Cold War. Alliance systems such as NATO and the Warsaw Pact were no longer valid because of the opening of Europe and the disintegration of the USSR. Arguably, the highest point of cooperation came with the signing of the Washington Charter in the spring of 1992. The document stated that neither state regarded the other as an adversary and both were committed to developing a partnership and friendship.

Cooperation also extended to other aspects, such as economic and military agreements. Russia joined the IMF and received significant loans in the 1990s. By 2006, Russia cleared its USSR-inherited debt to the Paris Club creditors ahead of schedule. Military cooperation extended to all levels. In January 1993, START II was signed, eliminating two-thirds of the nuclear arsenals for both countries. The two states cooperated in providing incentives to Belarus, Kazakhstan, and Ukraine to transfer their strategic nuclear weapons for safekeeping in Russia in return for economic incentives and territorial security guarantees. The latest manifestation in this sphere was the creation of a NATO-Russia Council in 2002 (superseded by the NATO-Russia Permanent Joint Council) as framework for consultation and cooperation on common interests.

Although each vital interest critically impacts Russia, the post-Soviet region is the most important and salient after the Cold War.[32] Kubicek concludes that "since 1991, its general pattern has swung from one of cooperation with the West to one of direct confrontation over issues such as Bosnia, NATO expansion, and Russia's assertion of a sphere of influence in other post-Soviet states."[33] This continuous shift, from cooperation to confrontation, follows closely to Vasquez's dictum: actual behavior, not just proclamations, will establish norms of peaceful cooperation.[34] In this sense, Russian policy and behavior negate the hope for rivalry termination, even during periods of cooperation, because they may be "just proclamations." We will next move on to a discussion of the specific issues at stake that increase the probability that rivalry remains ongoing. In this analysis, we follow the issue-based approach in our case study examination.[35]

Regional Issues of Rivalry: Post-Soviet Space

Over the past decade, post-Soviet space has continued to be a domain where US and Russian interests clashed either over policy (e.g., human rights, democratization, energy) or the territorial integrity of third-party states (e.g., Moldova, Georgia, Ukraine). Russians believe that they have a special role in this region for at least three reasons: the previous Soviet relationship, national security interests, and a Russian-speaking population resettled there during the Soviet era. Russian policy still claims the right to intervene in the affairs of these states, subsequently taking any intervention attempts by the West as encroachments on its own sovereignty.

The importance of the region cannot be understated as it provides buffer zones for Russia. During the Cold War, the Soviet Union established the Warsaw Treaty Organization as a countermeasure to NATO; it also controlled Eastern European governments to bolster the effectiveness of the buffer zones against the so-called "capitalist encirclement." Russian understanding of conflict vis-à-vis European countries was shaped by its historical experience. Therefore, it has always been imperative for Russian foreign policy to maintain defense and diplomatic relations with its neighboring states in order to establish zones of control. However, Russia's intent to keep the Soviet border framework alive—and indirectly the buffer zones—was hampered when NATO admitted new members in 1999 and 2004. In essence, this brought Cold War foes border to border.

Russia's desire for control of former Soviet republics increases the probability of conflict. Russia desires to control the region for defense and prestige purposes, while the US seeks economic liberalization, political pluralism, and more recently, an intensified security relationship based around the Missile Shield project and energy issues. In the 1990s, Moscow lacked the real capabilities to support the aggressive posturing of its foreign policy. However, the petrodollar boom under Putin has filled the government's coffers with funds, confidence, and consequently the ability to assert its interests. US policy of promoting democracy and Russia's policy on buffer zones come into direct conflict in post-Soviet space, and the rivalry was revived during its dying breath because of these issue conflicts, particularly in the Caucasus region.

Perhaps the most irritating foreign policy stance that the Western governments have taken since Putin's ascension to power is its efforts to influence elections in post-Soviet space, as Nygren argues.[36] The

Color Revolutions in Georgia (the Rose Revolution of 2003), Ukraine (the Orange Revolution of 2004–05), and Kyrgyzstan (the Tulip Revolution of 2005) were all relatively peaceful protests of supposed election fraud. All revolutions involved the public outrage over the pro-Russian political coalitions supposedly fixing the vote count for their easy victories while Western election observers cried foul and announced that the pro-Western political coalitions were the clear victors. These public demonstrations led to new elections being held where the pro-Western Mikhail Saakishvili and his party won in Georgia, Viktor Yushchenko's coalition won in Ukraine, and Kurmanbek Bakiyev in Kyrgyzstan.[37] Russia accused the United States and the West of meddling in sovereign states' elections and influencing the outcomes, while the West fired back and accused Russia of using its power in the region to prop up corrupt yet pro-Russian regimes. Russia has still not gotten over these events, and it even went as far as to accuse the United States of meddling in its own elections in 2011 and 2012.[38]

The Russian government also supports the breakaway regions of Abkhazia and South Ossetia. Both are *de jure* parts of Georgia but *de facto* autonomous republics. As of December 2009, the international status of the two territories remains unclear in the aftermath of the 2008 Georgian-Russian War.[39] Russia has recognized Abkhazia and South Ossetia as independent states, while the rest of the international community has been unwilling to support this position, including its allies in the Shanghai Cooperation Organization. Nagorno-Karabakh is yet another self-declared republic while officially part of Azerbaijan. Accompanying these ethno-nationalist territory-based conflicts is the economic conflict in Caucasus region. The region is valuable to the West because of its oil wealth, which allows for diversification of energy supplies. The Baku-Tbilisi-Ceyhan (BTC) pipeline carries oil from the Caspian to the Mediterranean Seas and runs through the territories of Azerbaijan, Georgia, and Turkey.[40] The US sees this pipeline partially located in post-Soviet space as a way to undermine Russia's tight grip on energy transport in the Caucasus region, while Russia sees its rival meddling in territory that is exclusively in Russia's sphere of influence. Perhaps the 2008 war on Georgia was a message to Georgia to not attempt assimilation with the West, but also a message to the US for it to "butt out" of Russia's former empire's affairs.

As the Russian economy bounced back from the 1998 financial crisis, this willingness to assert itself has continued to increase not only within the post-Soviet area but also gradually on a larger scale (e.g., the joint French-German-Russian opposition to the 2003 Iraq War). After the war between Georgia and Russia in August 2008, the US unambiguously sided with Georgia, prompting a further increase in tensions with Russia. Russian leaders then accused their rival of abetting Georgia and of encouraging Mikheil Saakashvili to attack the breakaway regions.[41] The partnership between the US and Georgia meant that Americans provided anti-terrorist training to Georgian forces; in return, Tbilisi sent 2,000 of its soldiers as part of the US mission in Iraq. Needless to say, the US found itself in an awkward situation of having to airlift Georgian troops from Iraq to their home country in order to counter an attack from Russian forces. As the War on Terror developed and Color Revolutions took place, the Caucasus area increased in its importance to both sides.

Another separatist enclave to note lying in Nygren's European Security Complex is Transdniester in Moldova.[42] The Russian-speaking Transdniestans have wanted political autonomy from the larger Romanian-dialect Moldova since the breakup of the Soviet Union. Russia has supported Transdniester's autonomy and has backed it up when Moldova has attempted to assimilate the region more forcibly.[43] This has kept this conflict frozen in time where Moldova does not have complete control over its borders and Russia has been able to coerce the country due to its energy dependence (more on this is discussed in Chapters 5 and 6). This issue is not a hotbed of contention between the US and Russia, although the State Department has showed support for Moldovan sovereignty from time to time.[44] However, if this standoff heats up in the near future, we could see the two rivals butting heads over the troubles of the country.

Another potential frozen conflict that could see the US and Russia on opposite sides of the fence is the standoff in Nagorno-Karabakh between Azerbaijan and Armenia. This standoff has basically remained at status quo for over a decade. Russia has generally shown more support for Armenia, although it has been careful not to go too far so as to not completely alienate Azerbaijan. The United States has generally had more cordial relations with Azerbaijan, although it made Armenia happy when Congress officially declared the Ottoman massacre of Armenians during World War I a genocide in 2008.[45] However, there has been more American support for Azerbaijan in energy projects, as the US has backed and financed the construction of the BTC oil pipeline as well as the natural gas–carrying Baku-Tbilisi-Erzurum pipeline, both of which circumvent Russia. This lessens the geopolitical hold Russia has on the region and does not sit well with the Russian foreign policy elite.[46] Therefore, if Nagorno-Karabakh turns from frozen to thawed, we could see US and Russian involvement where Russia will back Armenia and the United States will back Azerbaijan.

The 2014 crisis in Ukraine is the most recent dispute in post-Soviet space that is fueling the US-Russia rivalry. What began with a series of protests in November 2013 because of then-President Viktor Yanukovych's decision to reject an economic deal with the EU in favor of a bailout and energy package from Russia, the discord in Ukraine has now evolved into a bloody civil war.[47] Since Yanukovych's ouster, Russia has annexed the Crimean Peninsula from Ukraine, has nullified the economic package, and is now being accused of arming and financing the ethnic Russian separatists in the east (these separatists seem to be the culprits behind the accidental shooting down of Malaysia Airlines flight 17 with a Russian surface-to-air missile).[48] Furthermore, the Ukrainian people have elected a new pro-Western government and president in Petro Poroshenko, who has restored Ukraine's economic ties to the EU. This has caused a great divide between Russia and the West, where the latter has been incrementally imposing economic sanctions because of the former's suspected involvement in the destabilization of Ukraine.[49] If this is the case, why is Russia willing to jeopardize its economy and relations with Western governments over a poor country such as Ukraine? The salience and identity factors of the theory of situational coercive diplomacy have explanatory power here. Ukraine is essential to the energy interests of Russia, and the origins of the Russian state are found in Ukrainian territory, making the probability of Russian coercive diplomacy against Ukraine very high.

Ukraine is not only the most salient state to Russia in terms of energy as it serves as the main energy pipeline go-between for Russian conglomerates and their EU customers. This is the topic of Chapters 5 and 6 and will be discussed in detail there. Ukraine is also of great importance to Russia's great power identity. The origins of the modern Russian state are traced back to the Kievian Rus clan, which is the region that contains the Ukrainian capital of Kiev.[50] Losing Ukraine to Western institutions and influence, therefore, is a threat to the great power identity of the modern Russian state.[51] Therefore, if more hard evidence connects Russian involvement in arming and funding the ethnic Russian separatists in Ukraine, then Moscow's intentions become clearer: Losing Ukraine to the influence of the United States and the West is unacceptable, and coercive measures must be taken to ensure that Ukraine remains in Russia's political orbit. "Putin then sought to maintain pressure on the West by fomenting a pro-Russian insurgency that flared up in Ukraine's mostly Russian-speaking industrial east in April, apparently hoping that a slow-burning conflict would help persuade the West to strike a compromise that would allow Russia to keep Ukraine in its orbit."[52] Furthermore, as the United States has imposed economic sanctions on Russia for its behavior with Ukraine, rivalry behavior is at play here as well. The US has escalated the rivalry with Russia, and now Moscow must react coercively. The future of Ukraine is at the mercy of the dynamics of the US-Russian rivalry. This 2014 crisis will not end with the two more powerful adversaries coming to military blows directly; but the Ukraine will suffer, as will other pawns in the global conflict.

American foreign policymakers have also shown disdain for the "passportization" of ethnic Russians living in the territories of former Soviet states.[53] Russia has granted passports and hence citizenship to most of these Russians. This has made the quagmires in South Ossetia, Abkhazia, Transdniester, Crimea, and Eastern Ukraine even more problematic for the host countries of Georgia, Moldova, and Ukraine. The United States sees this as an infringement upon state sovereignty and not as a constructive way to resolve conflict. Linking this issue to other issues of the former Soviet Union could only exacerbate US-Russian relations and heat up the resurgent rivalry as well.

Interventions and Rivalry

Issues in the Slavic region during the post-Cold War era had the highest potential for conflict between Russia and the United States, which is alluded to later in this chapter. Kubicek asserts that the Bosnian conflict showed that a pro-Western foreign policy would be contested within Russia and would provide a favorable environment for emergence of nationalist leaders.[54] Starting in 1992, Russian leaders actively worked with the UN to solve the Bosnian problem. As the war dragged on, Western allies came to support the Croatian-Muslim population. This policy was detested by Russian nationalists, who "complained that Russia had kowtowed to the West and abandoned its brother Orthodox Serbs."[55] When the Serbs shelled a Sarajevo marketplace in 1994, Russian leaders succeeded in bringing an end to the attack and convinced Western forces not to retaliate. After Serbian attacks on UN-administered territory, Srebrenica and Žepa, NATO bombed Serbian forces. The Dayton Accords, which put an end to the Bosnian War, were negotiated with minimal Russian consultation. From the war, the Russian elites learned that the West was not willing to directly fight Russia over Serbia. From NATO's point of view, the initial policy of conceding to the Russians failed when the Serbs continued to attack UN-controlled territory. Therefore, NATO decided that a more assertive bargaining stance would have to be undertaken with the Serbs and Russia to prevent future attacks.

On 24 March 1999, NATO began Operation Allied Force with the objective to protect the Albanian population in Kosovo from Serbian attack and genocide.[56] Russian leaders viewed this event as both an aggressive maneuver against a sovereign state and as an action unsanctioned by the UN Security Council. Arguably, the highest potential for conflict between NATO and Russian forces was at the airport in Pristina when the latter's airborne troops took hold of the facility. The tension was defused by convincing Hungary and Romania to revoke fly-over rights for Russian transport planes with reinforcements.[57]

At the time, Putin stated that it was "inadmissible, under the slogan of so-called humanitarian intervention, to cancel out such basic principles of international law as sovereignty and territorial integrity of states."[58] Undoubtedly, the idea of a greater Slavic brotherhood motivated Russian action in this case.[59] The assumption of Western guilt during the Kosovo affair and trust in Serbian actions could only have been facilitated by historical learning where Russia felt the need to come to the aid of its allies when attacked by the West. Mendeloff notes "Western claims of Serb 'ethnic cleansing' against Kosovar Albanians were dismissed out of hand, while the West was labeled as an unprincipled 'aggressor' bent on the destruction of Yugoslavia, and even Russia itself."[60] In the context of a continued rivalry, the events during the Kosovo conflict only reinforced these views and exacerbated tensions between the two states. Ironically, this war left the final status of Kosovo unresolved for nine years. Backed by the US and the EU, Kosovo declared independence in 2008. A few months later, this event was used as an excuse by Russia to recognize the independence of Abkhazia and South Ossetia.

More recently, Russia has directly opposed US and NATO opposition to the status quo regimes in Libya and Syria. As the US and the West see Libyan President Muammar Gaddafi repressing and killing his own people opposed to his regime, the pro-government Russian online mouthpiece *Pravda* suggests the dictator is benevolent and paints the opposition as terrorists armed by the evil organization that is NATO.[61] It is NATO, not Gaddafi, that is killing the Libyan people, and references are given to NATO intervention in the Balkans, Iraq, and Afghanistan as similar comparisons. Russia also perceives this as a strategic move by the US to block Russian energy giants Gazprom and Rosneft from access to Libyan oil and natural gas. Libya is merely a part of a bigger US-Russian "secret oil war," where energy interests of the rivals are competing worldwide.[62] Russia believes it was tricked into supporting the UN resolution to protect Libyan civilians, as it did not realize that NATO's involvement would be so entrenched. This prompted Russia to stand firm on Syria.

In Syria, *Pravda* portrays the US and the West as supporting "terrorists," referring to the opposition that took to the streets in 2011 demanding reform and regime change.[63] The row in Syria has since expanded into a full-fledged civil war. Russia legitimizes the violence used by the Syrian government because of this terrorist portrayal. Russia supplies Syria with large amounts of arms and fears losing a valued customer.[64] Only when a proposed UN Security Council vote, which Russia threatened to veto, to decide what to do about the Syrian government's actions and the death toll of civilians at the hands of Syrian President Assad become too great to justify did the Russian Foreign Ministry condemn the latter's methods. However, Russia, along with China, still believes that calls for Assad to step down are unwarranted and that the West as well as the UN should stay out of Syria's domestic affairs, even in the face of Assad potentially breaking a longstanding international norm and taboo by using chemical weapons.

There are two other critical issues to Russia that have to do with Syria: Syria is one of Russia's best customers in the purchasing of arms and weapons technologies, and the only naval base outside of the borders of the former Soviet Union is at the Syrian port of Tartus.[65] The loss of this base would be a blow to Russia's resurgent attempts at reestablishing itself as a global power. However, due to the international backlash of Russia's continued sales of arms to the Assad regime in light of the escalating bloodshed, Russia has since announced that it will not enter into any new arms deals with Syria until the crisis is resolved.[66] However, Russia, as well as China, remains steadfast in its assertion that the conflict should be resolved internally, and it is vehemently opposed to international military intervention.[67]

Alliance Systems

The ongoing NATO enlargement policy troubles Russia for a number of reasons. The alliance was originally created in 1949 to counter the Soviet threat. With the end of the Cold War, the organization had to redefine its mission if it was to avoid replicating the UN and its peacekeeping operations. The US envisioned NATO as an instrument to stabilize former Soviet territories during their transition to democracy while Russia envisioned a new NATO as either cooperating equally with its armed forces or disintegrating completely. Simultaneous to mission reformulation, NATO continued to expand. Russett and Stam write, "whatever Westerners may say, that kind of expansion is directed against at least a hypothetical danger from Russia."[68] In 1999, the Czech Republic,

Hungary, and Poland entered the alliance. The inclusion of these countries proved troubling for Russia, especially since they had received assurances during the reunification of Germany that NATO expansion would stand idle. The failure of the US to keep this promise only reinforced the idea that the US could not be trusted and that it was really out to encircle Russia.

NATO's role in the world increased following 9/11. NATO invoked Article Five, which directed that "an armed attack against one or more [members] in Europe or North America shall be considered an attack against them all."[69] By summer of 2003, NATO took command of an International Security Assistance Force in Afghanistan. The largest expansion took place in 2004 as seven new members from the Baltic region and Eastern Europe joined NATO. Russian territory now shared a border with three NATO states: Poland, Latvia, and Estonia. Moscow's leaders vocally opposed these enlargements. Sergei Lavrov, Russian Foreign Minister, told the Council of Europe that "NATO's steady enlargement perpetuated an old 'bloc' approach to resolving international problems" and called on the West not to erect new dividing lines.[70] Lavrov's criticism is consistent with the Foreign Policy Concept of the Russian Federation, which states: "Russia maintains a negative view towards NATO enlargement" (part IV, paragraph 14). Faced with an enlarged North Atlantic alliance, Russian leaders had two possible courses of action. Russia could either form a counter alliance or build up its own military capabilities. Both of these are traditional strategies among *realpolitik* practitioners.[71]

Russia has mimicked NATO's new mission statement for the CIS, or Commonwealth of Independent States, where an attack on one CIS member jeopardizes the security of all CIS members and is a clear violation of Russia's vital interests and considered an act of war.[72] At one time or another, the CIS consisted of all former Soviet republics except Estonia, Latvia, and Lithuania. After the 2008 Russo-Georgian war, Georgia withdrew from the organization, which also prompted Ukraine and Turkmenistan to postpone officially rejoining the CIS. Furthermore, in 2009, the six countries of Armenia, Azerbaijan, Moldova, Georgia, Ukraine, and Belarus signed on to the EU-initiated Eastern Partnership, which was a huge blow to Russian control over its former vassal states. This new economic alliance with the West could only further exacerbate Russia's feelings towards the West, including the United States.

Along with six major CIS states either withdrawing from the organization or joining a pro-Western economic alliance, many CIS states have not renewed the 1994 security treaty of the CIS, the Collective Security

Treaty Organization (CSTO), which at its largest consisted of Russia, Armenia, Kazakhstan, Kyrgyzstan, Tajikistan, Uzbekistan, Azerbaijan, Georgia, and Belarus. In 1999, Azerbaijan, Georgia, and Uzbekistan refused to renew the treaty and instead formed, along with Moldova and Ukraine, a pro-US alliance known as GUAM (Uzbekistan has shifted its position between GUAM and CSTO several times). Russia has had rocky relationships with these countries due to apparent grudges against the US for its meddling in perceived exclusive Russian space.[73]

The United States' policy supporting NATO enlargement and democratic movements in post-communist states (Color Revolutions) are examples of denying potential influence to a rival. In addition, support for Kosovo's independence, despite Russian assurances that such action would result in their own push for independence of contested territories, demonstrates that the United States will do what is in its own interests regardless of Moscow's wishes. Ignoring Russia's views is dangerous due to its hold on energy resources, veto power in the UN Security Council, and influence in Iran and North Korea.

Missile Defense Systems and Military Buildups

Russia's sale of arms and dual use of technologies to countries the US considers unfriendly or outright hostile is another area of conflict for the dyad. Russia continues to supply arms to states like Venezuela, Syria, Iran, India, and China. The Russian military industry also signed contracts with Malaysia and Indonesia in 2003, with Venezuela in 2006, and it delivered anti-aircraft missiles to Iran in early 2007.[74] Prior to the American invasion of Iraq, Saddam's regime was also a customer of the Russian military industrial complex. Control of dangerous arms transfers is of vital interest to the United States and remains another facet of rivalry with Russia.

In its turn, the US contributed to the perpetuation of the rivalry by insisting on continuing with the missile shield project in Eastern Europe up until Obama canceled most of the program in late 2009. The American side argued that the purpose of the shield was to counter an emerging threat from rogue states (Iran and North Korea). The US further claimed that unarmed missile interceptors that use the kinetic force to destroy their target present no danger to Russia's nuclear deterrent. Despite this, Russia was vociferously opposed. Russia objected to increasing US involvement in Eastern Europe so close to its borders and suspected that the radar base in the Czech Republic could be used for spying on Russia.

In May 2010, the US installed several Patriot missiles at a Polish base, only a few miles from the Russian enclave of Kaliningrad. The US reiterated that the missiles were to defend NATO countries from possible missile attacks from Iran or North Korea. However, as the missiles are so close to Russian soil, it has caused alarm in Moscow.[75] Russia has questioned the strategic placement of the missiles, as a missile defense shield from Iran would better be located in NATO member-state Turkey, while longstanding American allies Japan and South Korea would be better locations for defense against North Korea.

The US has also made arrangements with Romania to install SM-3 interceptor missiles on Romanian territory, the same missiles that Obama had decided not to deploy to Poland. Romania, although closer to Iran, is also very close to the naval base of Russia's Black Sea Fleet in Crimea.[76] In June 2011, a US cruiser equipped with the Aegis ballistic missile defense system appeared in the Black Sea, much to Russia's dismay. There have also been discussions between the US and Bulgaria for deployment of missiles; however, these talks have been scrapped for now. All of these developments have flexed US military muscle in Russia's backyard.[77] Alarmed, Russia has denounced the US's Black Sea actions.

The downing of Malaysia Airlines flight 17 in July 2014 shows the importance of surface-to-air missiles, such as the Russian Buks, getting into the wrong hands as an international security issue. When these sophisticated weapons get into the hands of separatists never supposed to have this access, disasters may happen. Securing these missiles rather than using them as indirect tools to perpetuate a rivalry is irresponsible for both the United States and Russia. Some in the media are now calling Russia a "rogue state" with no regard for international law or regional stability.[78] The United States should also know better than to install missile defense systems so close to Russia's borders, even if they are meant to shield from attack from the Middle East. These most recent issues between these longstanding rivals are disrupting regional stability. Perhaps another factor in our theory of situational coercive diplomacy is also fueling Russian aggressive policy where it sees its natural interests threatened: the factor of elite and public opinion as domestic sources of rivalry.

Elite and Public Opinion: Domestic Sources of Rivalry and Russian Coercive Diplomacy

Holsti suggests that public opinion in the post–Cold War world is as important as ever in directing the actions of foreign policymakers.[79] Furthermore, he concludes that with the end of the superpower rivalry and the beginning of a new era of US-Russian relations, there will be fewer major power confrontations, and therefore we may have entered an era where public opinion plays a more autonomous role in affecting foreign policy decisions by state leaders.[80] Crises and confrontations internationally, especially between rivals, provide a setting for which there are ample opportunities for elite manipulation of the masses. These types of issues between rivals are resolved over a longer period of time and provide greater opportunities for the public and media to play a larger role in foreign policy.[81] Public opinion in the context of crises or confrontation also tends to be more resistant to claims that the "need for secrecy, flexibility, and speed of action make it both necessary and legitimate for the executive to have a relatively free hand."[82]

A new relationship between public opinion and policy is at the heart of the Russian and American public, serving as an important force for decisions of their leaders regarding these adversaries' interactions. We agree with this view; therefore, American and Russian public opinion polls are an important reason why the US-Russian rivalry has yet to terminate. Elites can use these sentiments to drive action, and these sentiments can drive elites to take action. Coercive diplomacy over issues regarding post-Soviet space in particular, we find, have widespread public support. This does not imply causation. We assert that public opinions do not dictate foreign policy decisions, but they do serve as a guide and affirmation of the decisions that are ultimately made.

Page and Shapiro stress that there is a difference between an individual's opinion changes on an issue, which can occur in a rather short period of time, and that of collective public opinion on an issue, which remains very static throughout time.[83] When "fluctuations" in collective public opinion occur, they are usually preceded by political shocks that can cause widespread opinion change. Foreign policy crises such as Vietnam, the Cuban Missile Crisis, and the recent economic recession, Page and Shapiro argue, can cause massive fluctuations in the collective American conscious, which the public learns from and usually does not revert to the pre-crisis opinion.[84] With the collapse of the Soviet Union, for example, the American public had an optimistic view about the future of relations with Russia. However, the opinions have gradually become more pessimistic with the actions of a more assertive Russia under the Putin Administration. This fluctuation of optimism has reverted back to mistrust of Russia, albeit with less consensus.[85]

The central premise of this section of the chapter is that coercive diplomacy is used by Russia in situations of rivalry or when the salience of the issue is high for public opinion because these issues can then be used by the leadership to precipitate action. Public opinion can also signal that a rivalry is ongoing, adding support to evidence presented in the first half of this chapter as well as in Chapter 6 that Russia maintains ongoing rivalries. Along with disagreements in post-Soviet space and over international issues, we find that the elites and public of the US and Russia still see each other's countries as enemies and not to be trusted. There is evidence that anti-American sentiment in Russia is stronger than the anti-Russian sentiment by the elites and public of the United States; however, anti-Russian sentiment exists nonetheless. This section develops the hypothesis that rivalries cannot terminate until the elites and public of the rival states in question discontinue their enemy perceptions. Domestic impulses in Russia and the United States can help push them to act internationally in either a coercive or accommodating manner. An examination of public opinion polls from both countries confirms this inclination.

Measuring Issue Salience: Elite and Public Opinion

Putnam's two-level game model can be used to illustrate the current state of affairs between Russia and the United States.[86] Putnam develops the notion of a two-level ratification game for international political processes. A leader or actor must first come to an agreement or decision about an action on the international level. The actor must then form a "win set" on the domestic stage to approve policy options. Failure to gain a "win set" will result in loss of internal support and possibly removal from office, which is even possible in a semi-authoritarian Russia. Gaining leverage and approval at both the international and domestic levels are critical for a leader trying to propose new policy options, including the termination of a rivalry relationship.

Public opinion becomes important in light of Putnam's two-level game theory. It can help provide the necessary "win set" needed at the domestic level. Although perhaps more difficult for Russians, the public has the ability to remove from office leaders whose policies do not satisfy them.[87] Mor finds that public support is important for the mobilization of society required under an intense international rivalry.[88] Public opinion can also serve as a restraint on leaders' attempts to continue a rivalry and may force reconciliation. Convergence of preferences between the public and elites may be a necessary condition for the continuation or termination of rivalries. It is also the case that while the principal actors (i.e., political leaders) learn how to respond to a crisis within a rivalrous relationship, the public also learns that a rivalry relationship is present and may become an obstacle for diplomatic solutions to the protracted conflict if this group is not satisfied.

Applying the two-level game approach to the Russia-US rivalry will show that attempts to end the rivalry in the years 1986 to 1993 did not succeed. To continue the retention of office, Boris Yeltsin and Vladimir Putin needed to create a win set at the domestic level that included those with negative opinions of the United States. Given that the nationalist leaders such as Aleksandr Lebed and Vladimir Zhirinovsky exhibited foreign policy preferences that included the continuation of the rivalry, Yeltsin adopted these attitudes to ensure his reelection and co-opt this position. Later, Putin's nostalgia for Soviet prestige and view that Russia must be recognized as a great power has encouraged continuing rivalry with the US. Recent concern with action in Libya plus Syria, and the forced rejection of American aid and cooperative agreements, only keeps them on this path. In retaliation for a decision by the American Congress to punish Russians who have committed human rights violations, the Russian Duma passed a total ban on American adoptions of Russian children. This is the behavior of rivals, not friends.

Through a process of learning, the Russia-US rivalry endures. Russian domestic forces, both elites and the mass public, still see the United States as a threat to the Russian sphere of influence. This sphere involves regional issues such as Serbian sovereignty and the status of Russian minorities in Baltic States, Georgia, and Ukraine. These issues, as well as other former Soviet areas in contention, have led to a conflict of interest with the United States.[89]

The most cogent work on rivalry termination and escalation has been presented by Colaresi in his treatment of the sources of rivalry persistence.[90] Colaresi's theory of "rivalry outbidding" suggests that a leader will have to continue a rivalry if the elites and mass public wish to continue the relationship, thus either elites or the public can overrule and outbid a leader. Structural changes can often have little impact on the course of the rivalry if the leader is not in a position to terminate a hostile relationship. Often, the leadership will use this public hostility to channel domestic power to his or her regime.

Colaresi's work points to the importance of domestic elite and public opinion in setting the issues at stake in a rivalry. While domestic politics critically impacts a rivalry situation, what exactly is the consequence of the public and elites in terms of ending a rivalry situation? We argue that without the recognition of the termination of the rivalry relationship and the settlement of the issues by the public and elites, a rivalry will not terminate. Domestic acceptance is important since leaders and regimes change and shift. When the public and elites do not agree with the goals of the leadership, they will only seek to reignite the rivalry once another transition of leadership occurs.

Using the issue-based approach as our guide, we posit that the issues that the US and Russia have differences over are translated into the opinions of the elites as well as the public in both countries. A positive feedback loop where the elites take cues from public opinion and the public takes cues from elite opinions and actions occurs, and the principal rivals continue to be "locked-in" as adversaries.[91] Salience of issues can be partially measured through public opinion. We posit that domestic impulses resonated through public opinion either encourage or discourage coercive diplomacy abroad. If there is enough consensus on an issue, the Russian government will perceive it to be of high salience and act coercively. If there is no consensus or little public interest in an issue, the Russian leadership will interpret the issue to be of low salience and not act upon it or seek cooperative negotiation over coercive diplomacy. This brings us to the hypotheses of this section:

> *H3: Elite and public opinions about the adversarial state are domestic sources of the continuation or termination of rivalries, as elites take cues from public opinion and vice versa.*
>
> *H4: The closer to unanimous that public opinion on international issues is, the more public opinion serves as cues to governments as to how salient an issue is, and this motivates governments to act coercively or seek cooperative outcomes.*

The remainder of this chapter will present and analyze public opinion polls taken in Russia and the United States since the ascension of Putin in 2000 to the present. The next section covers Russian public opinion of the United States, followed by American public opinion about Russia. The issues covered include alliances, weapons placements and sales, and military interventions. We utilized just about every public opinion poll we were aware of, avoiding selecting information based on its support of our hypotheses in favor of a search for the true state of opinion on Russia and its actions.

There have been a number of public opinion studies regarding foreign policy viewpoints of average Russian voters. Here we present several time-series and cross-sectional public opinion polls that poll Russians and Americans. Polls were taken by a variety of public opinion centers in the United States, Great Britain, and Russia and include Pew, Gallup, BBC, WPO, Russian Public Opinion Center, and Levada.[92] The tables and figures show that Americans and Russians are still very suspicious of one another.

We begin with Russian opinions of the United States and American foreign policy. Figure 3.1 represents the Russian public's overall views of the US over time. These fluctuations in opinions about the United States suggest that domestic public opinion about international issues in Russia are good indicators as to how Moscow will deal with the United States at a particular point in time. Salience of issues with the United States, therefore, should also fluctuate. The spike in favorability of the United States coincides with Putin's pledge to support American efforts to combat terrorism.[93] The dramatic drop in favorability towards the US in 2003 can be attributed to the global unpopularity of Bush's invasion of Iraq. Putin was a vehement opponent of the US invasion.[94] The slight drop in favorability in early 2006 is likely due to the Cheney-Putin dialogue on Russian human rights abuses.[95] The huge unfavorable spike towards the US in 2008 can be attributed to a combination of the US's stance on Russia's war with Georgia as well as the stock market crash that destabilized the global economy.[96] The opinions of the Russian public and Russian policymakers, therefore, are sound determinants of whether or not elites will use coercive diplomacy or accommodationist strategies with their longtime rival.

Table 3.1 shows that Russians also believe that the United States coerces Russia into bending its will to an American point of view. The

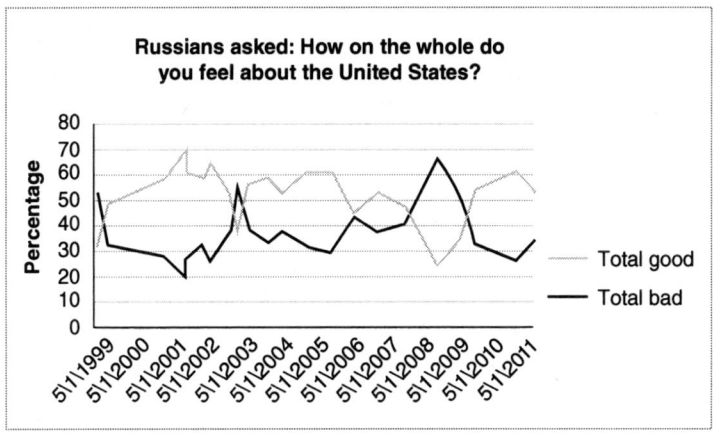

Figure 3.1 Russian opinion of the United States over time
Source of data: Levada Center. *Levada.ru*. Dates: As noted.

Table 3.1 Russian opinion of US treatment of Russia

Respondents asked: Do you think the US more often treats Russia fairly, OR abuses its greater power to make us do what the US wants?

	2008	2009
Fairly	19%	12
Unfairly	69	75
No opinion	12	13

Source of data: WPO/Levada Center. *Levada.ru.* Dates: May 22–26, 2009.

inferior feeling of lost superpower status Russia has endured since 1991 is growing based on the results of these recent polls. Russian nationalism as well as the Russian psyche has yet to let go of the animosity felt towards the United States. After 50 years of being taught that the United States is Russia's number-one foe, the process of unlearning these views has yet to proliferate. Russians are aware that the United States won the Cold War, that Russia is no longer equal in power to their Cold War rival, and that the US now has a freer hand in global affairs.[97] This gives Russians the notion that Russia is no longer respected or treated as an equal with the United States, and this resentment is reflected in the public opinion polls in Table 3.1. Russians still see the United States as their enemy, which is a powerful domestic source of the continuation of a rivalry.

Table 3.2 uncovers which five countries Russians believe are their top enemies, asking respondents to list who they think is unfriendly or hostile to Russia. The top two countries over time are the countries that are embroiled in a rivalry with Russia: Georgia and the United States.[98] We posit that this is no accident, as issues with rivals tend to be more salient to states. The country cited the most, especially after the 2008 conflict with Russia, is Georgia. Even before the five-day conflict, Georgia was considered the most hostile state in Russia's Near Abroad. This powerful domestic indicator could partially explain why Russia was willing to resort to armed conflict with the former Soviet state, even in the face of possible American intervention.[99] This strong anti-Georgian consensus sealed the fate of the Georgian military in August 2008.

The second most hostile state to Russia, according to the Russian public, is the United States. Russian perceptions of the United States seem to have remained unchanged even in the face of the end of the Cold War. Other states listed in Table 3.3 are states that have either integrated with the West (the three Baltic States) or have sought Western help in the face of Russian coercive foreign policy (Ukraine).[100] The Baltic States

Table 3.2 Russian opinion of states it considers unfriendly or hostile

Respondents asked: Name five countries that you would consider unfriendly or hostile to Russia.

	2005	2006	2007	2009
Georgia	38%	44	46	62
United States	23	37	35	45
Ukraine	13	27	23	41
Latvia	49	46	36	35
Lithuania	42	42	32	35
Estonia	32	28	60	30

Source of data: Levada Analytical Center's Annual Russian Public Opinion Yearbook 2009. http://en.d7154.agava.net/sites/en.d7154.agava.net/files/Levada2009Eng.pdf.

have joined NATO and the EU. States that ally themselves with Russia's longstanding enemy, the US, are also perceived as hostile towards Russia.[101] This shows the importance of the domestic sources of rivalries: the friend of my enemy is no longer my friend. Former Soviet states that veer towards the West will be met with coercive Russian diplomacy in one form or another. This topic is discussed more in Chapter 5.

It is clear through the analysis of Russian elite and public opinion that Russians still have strong negative feelings towards the United States. The American issue was and still is highly salient to most Russians. These domestic factors make it difficult for Russia to wrest itself out of conflicting and rivalrous foreign policy practices with its longtime enemy. For Russia, the US-Russian rivalry is alive and well, and the mistrust of the world's sole superpower is strong. We now turn to how the American public perceives Russia. Is their animosity as strong as the Russian public's?

Table 3.3 shows the favorable/unfavorable index of the American public's overall view of Russia. Americans are generally split on their opinion of Russia, as the number of those who favor Russia roughly equals those who do not. However, as the years have passed, those who see Russia viewed as unfavorable now outnumber those who favor Russia. Explanations for this growing American animosity are plentiful. President Barack Obama's attempts at resetting relations with Russia have, for the most part, failed.[102] The Russian-Georgian war in 2008 has increased this growing American animosity since the US is a supporter of Georgia.[103] Russian support for the status quo governments in Libya and Syria have left the American public bewildered about Moscow's support for perceived despotic leaders.[104] Lastly, the worldwide publicity of suspected

Table 3.3 American opinion of Russia

Respondents asked: Do you have a favorable or unfavorable view of Russia?

	2007	2009	2010	2011	2012
Very favorable	4%	7	7	8	5
Somewhat favorable	40	36	42	41	32
Somewhat unfavorable	24	27	24	22	27
Very unfavorable	11	12	8	10	13
No opinion	21	18	19	19	24

Source of data: Pew Global Attitudes Project Key Indicators Database. http://www.pewglobal.org/database/?indicator=27&survey=14&response=Unfavorable&mode=chart.

fraud in the recent Russian elections and Putin's harsh anti-American response will only continue to motivate an increase in Americans' unfavorable views of Russia. With growing animosities about Russia after a period of more favorable views, the domestic sources of rivalry on the US side are rebounding, indicating that as issues with Russia become more salient, US policymakers will take stronger stances against coercive Russian foreign policy practices.

Figure 3.2 depicts Americans' growing mistrust and overall negative views of Russia and the foreign policy practices of the Putin regime. It seems that as authoritarianism has gradually increased in Russia, so has America's unfavorable opinion of the state. Focusing events, such as the invasion of Georgia or the Estonian cyber incidents, do not seem to determine public opinion of Russia in the United States, as the up-and-down spikes seen in the Russian opinion figure are not manifest in the American figure. Overall, these elite and public opinion studies show that Russians and Americans are not ready to trust each other, and each public views the other's government with suspicion and reservations. These types of opinions, if maintained, can only perpetuate mutual US and Russian suspicion and hostility. A more recent 2013 poll finds that over half of the American public now sees Russia as an enemy or hostile.[105] With this increased opinion of animosity, American and Russian relations could return to the zero-sum, dangerous games akin to those of the Cold War. The chart can be interpreted as parallel to the growing salience of the Russian issue in general among Americans.

When it comes to what Americans think about Russian foreign policy and its use of military power, Table 3.4 shows that there is much more disapproval than approval of how Russia conducts its international affairs. Many Americans remember Cold War times where military standoffs and coercive Russian diplomacy was the norm.[106] Russian

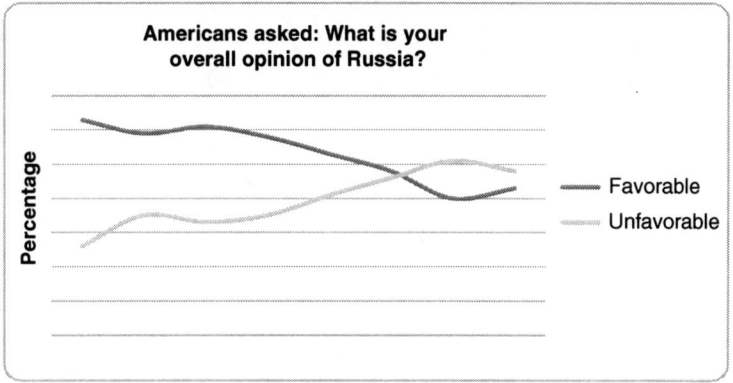

Figure 3.2 American overall views of Russia
Source of data: Gallup. *Gallup.com*. Dates: As noted.

Table 3.4 American opinion on Russian foreign policy

	Russian foreign policy influence	Opinion of Russian use of military power
Positive	40%	24
Negative	53	68
No opinion	7	8

Source of data: World Public Opinion (May 2006). http://www.worldpublicopinion.org/pipa/articles/views_on_countriesregions_bt/200.php? Lb=breu&pnt=200&nid=&id=.

opposition to American national interests continues to this day, thus we find this reflected in the public opinion polls taken in 2006. This is when more Americans view Russia as favorable than unfavorable as indicated in Table 3.2; however, American perceptions of how Russia conducts its foreign policy seems to remain unchanged in regards to Cold War–era behavior. Although more Americans may have thought the Russian regime to be more favorable than the Soviet regime, the opinion of Russia's Soviet-era foreign policy behavior remains untrustworthy and negative. This issue's salience has remained high.

Evidence uncovered in the preceding polls confirms the hypotheses of this section and also strengthens the arguments made in the first half of this chapter. Russian public opinion overwhelmingly sees the United States as an adversary and as a roadblock to Russian national interests and foreign policy objectives. The issue of Russia for Americans, therefore, is growing in salience. Russian domestic sources of rivalry confirm that the United States and Russia are still embroiled in a rivalry, albeit a regional and issue-based rivalry rather than the global ideological rivalry of old. American public opinion gave Russia the benefit of the doubt for a short interval with the collapse of the Soviet Union, as more Americans have seen Russia as favorable rather than unfavorable until very recently. Evidence of Russian coercive foreign policy, using public opinion as a measurement, is further confirmed with the issue of alliances and with states with which Russia should cooperate.

Russian Public Opinion on Cooperation and Alliances

Public opinion can also indicate whether or not states should cooperate or even form alliances. Russia, it seems, is willing to cooperate with any state that openly opposes the US, especially states of the former Soviet Union. Regarding Russian opinion of the NATO alliance, the poll results of Table 3.5 show that there is some disconnect among Russians and particularly Germany, a key member in the traditionally anti-Russian alliance. Sixty-three percent of Russians see NATO as hostile to the Russian state, even when the leaders of the alliance assert that it is no longer directed at protection from possible Russian aggression. We can only conclude that the reason for such anti-NATO feelings about an organization whose members Russians also seek cooperation with and consider one powerful member an ally is because Russians see NATO as an outcropping of the national interests of the United States.[107] NATO is not European, but rather American, in the eyes of most Russians. NATO "aggression," therefore, is American aggression, and the actions and expansion of the alliance is of purely American design.

Table 3.6 shows the opinions of Russians on whether NATO members have a reason to fear Russia. Perhaps a loaded question, the results indicate the amount of Russian pride and nationalism the public still hold for their country. Nearly two-thirds of Russians over the entire time span of the polling believe that NATO members would be wise not to cross Russia. This overwhelming majority of Russians who believe that NATO should fear Russia can be explained once again by the attributes of the domestic sources of the US-Russia rivalry and why Russia takes such a hardline stance on the organization and shows that the NATO issue is very salient to Russia.

The possibility of two former integral republics of the USSR joining NATO is seen as a threat for most Russians, as evidenced in Table 3.7. More recent polls show that about three-fourths of Russians see the ascension of Ukraine and Georgia into the American-led alliance as a threat of some kind, with most having this view perceiving it as a serious threat. Russian energy coercion against Ukraine and military intervention against Georgia allude to the salience of this issue, which most Russians deem a threat to their country's national interest.[108] Georgia took action against its Russian-backed separatist enclave of South Ossetia assuming that American friendship would deter a Russian military response.[109] They underestimated how salient this issue was to Russia. Similarly, the Russo-Ukrainian gas crises of 2006 and 2009 happened under pro-Western "Orange Revolution" rule, and perhaps Ukraine believed that it could take a stronger stance against Russian energy coercion.[110] This was also a miscalculation, and natural gas supplies were shut off to Ukraine and parts of Europe in the dead of winter. Meddling of the West in post-Soviet space is taken very seriously by Russians and their government.

Russian public opinion on alliances, states Russia should cooperate with, and the future of post-Soviet space uncover the desire of most Russians to regain Russia's lost Soviet power. Russians have been standing up to the United States and NATO for over 60 years, and this remains largely unchanged. Pitting Western European states against the United States in an attempt to bring them closer to cooperation with Russia,

Table 3.5 Russian opinion on the motives and purpose of NATO

Respondents asked: The leaders of the NATO coalition assert that the coalition is not directed against Russia. What do you make of this?

Yes, this is true	13%
No, the NATO coalition is still hostile to and allied against Russia	63
No opinion	25

Source of data: Levada Analytical Center's Annual Russian Public Opinion Yearbook 2009. August 2009. http://en.d7154.agava.net/sites/en.d7154.agava.net/files/Levada2009Eng.pdf.

Table 3.6 Russian opinion on whether NATO members in Europe should fear Russia

	1999	2000	2001	2002	2005	2006	2007	2008	2009
Definitely yes	27%	16	25	14	15	22	14	23	22
Yes, rather than no	36	38	33	42	43	40	35	39	40
No, rather than yes	22	25	22	23	26	22	27	21	21
Definitely no	7	7	8	7	8	7	8	8	6
No opinion	10	14	13	14	9	10	16	10	12

Source of data: Levada Analytical Center's Annual Russian Public Opinion Yearbook 2009. http://en.d7154.agava.net/sites/en.d7154.agava.net/files/Levada2009Eng.pdf.

Table 3.7 Russian opinion of Ukraine and Georgia joining NATO

	Ukraine 2006	Georgia 2006	Ukraine 2007	Georgia 2007	Ukraine 2008	Georgia 2008	Ukraine 2009	Georgia 2009
Serious threat	25%	27	33	36	36	45	34	38
Some threat	28	27	28	37	32	29	36	32
Little threat	15	15	14	14	12	10	13	15
No threat	18	16	10	9	8	7	8	7
No opinion	15	16	15	14	11	10	9	8

Source of data: Levada Analytical Center's Annual Russian Public Opinion Yearbook 2009. http://en.d7154.agava.net/sites/en.d7154.agava.net/files/Levada2009Eng.pdf.

and keeping the West and Western interests out of post-Soviet space, are other desires that the majority of Russians have. This is all too clear according to our hypotheses on public opinion affecting issue salience and Russian foreign policy behavior. This is further confirmed in the next section, which tackles Russian public opinion on US placement of anti-ballistic missiles in Eastern Europe, a region that used to be in the Soviet sphere of influence.

Russian Public Opinion of US Anti-Ballistic Missiles in Europe

Russia's official government stance on the placing of American anti-ballistic missiles (ABMs) has been adversarial. Although America's official motives for the placement of these systems so close to Russian borders is for protection of Europe against a possible Iranian nuclear strike, Russia sees this logic as flawed, and the only rational reason for placement in Poland and the Czech Republic is to deter Russia and to gain geopolitical advantage over Moscow in Eastern Europe. Russia reasons that ABMs to deter and protect from Iran would be better placed in Turkey or even Iraq.[111] Placement of ABMs in Eastern Europe equates to one motive in the minds of Russian leadership: an American power play at the expense of Russia.[112]

These motives are confirmed in the results of a Russian public opinion poll taken in 2009 and presented in Table 3.8. Fifty-seven percent of Russians see the placement of ABMs in Russia's former backyard as either an American desire to secure military advantage over Russia or to exert geopolitical pressure on Russia. A further 18 percent of Russians believe

Table 3.8 Russian opinion about Russian military intervention in Georgia in 2008

Russian authorities did their best to prevent escalation of conflict and slaughter	70%
Russian authorities got involved in a conflict that will backfire on Russia	16
Russian authorities instigated conflict between Georgia and South Ossetia	4
No opinion	9

Source of data: Levada Analytical Center's Annual Russian Public Opinion Yearbook 2009. September 2008. http://en.d7154.agava.net/sites/en.d7154.agava.net/files/Levada2009Eng.pdf.

the ABMs are to protect Europe from a possible Russian nuclear or conventional attack, while only 8 percent believe the actual stated American motive that the systems are for protection from a nuclear strike by Iran. These opinions by the Russian public reflect those of the Russian elite and show distinct Cold War–type paranoia and brinksmanship thinking.[113] This negative view of American motives uncovers more evidence of this section's hypotheses, as domestic public opinion on this issue about the United States confirms the continued existence of the rivalry.

The public opinion about American ABMs in Eastern Europe indicates the high salience of the issue of American power and its growing proximity to Russian borders. These opinions reflect the strong stance Russia has taken against the United States on this issue and is a probable motivating factor behind President Obama's decisions to scrap/postpone the installation of these sites. The Obama Administration found it not to be worth the headache and risk of coercive maneuvers from Russia. Russia's assertiveness with the United States and the West does not stop with ABMs, nor is the salience according to public opinion much different from Russian positions on recent military interventions involving itself or its interests, the United States, and NATO.

Russian Public Opinion on Military Interventions

Russia and the West have not seen eye to eye on many military interventions and foreign policy crises since the fall of the Soviet Union. It seems that where the United States is involved, Russia is there to oppose, and vice versa. Denying the other side's foreign policy aims is a classic example of rivalry behavior.[114] There have been brief periods of cooperation between Russia and the US, particularly after 9/11, when the Russians approved of troop deployments and air bases in Central Asian post-Soviet space. However, looking at Russian public opinion polls about interventions and crises that include Ukraine, Georgia, Bosnia, Kosovo, Iraq, Libya, and Syria, we find that the perceptions of the Russian public coincide with the foreign policy behavior of the Russian government, giving it the confidence to stand up and oppose American interests in these recent conflicts. With regards to the civil strife in Ukraine, a June 2014 *Levada* poll finds that 64 percent of Russians are in favor of their government arming, funding, and supporting the ethnic Russian separatists in Eastern Ukraine, with only 17 percent opposed.[115] Although Putin has denied to the international community the Russian government's part in perpetuating the bloody conflict, it seems that any domestic opposition to the president making it official would be minimal. Public support for Russian coercion in post-Soviet space does not stop with Ukraine; there was also widespread support for the 2008 military skirmish with Georgia.

The five-day conflict between Russia and Georgia over the disputed region of South Ossetia saw widespread support in Russia and the same amount of condemnation throughout the West.[116] Looking at Table 3.9, a poll taken one month after the conflict demonstrates that 70 percent of Russians believe that the Russian intervention was just, and its main purpose was to protect civilians from the Georgian onslaught in South Ossetia. The United States was caught off guard, the Georgians were shocked, and Russia was able to send a clear message to the West about its seriousness about keeping the Near Abroad solely under Moscow's sphere of influence.[117]

Other polls about the Georgian intervention taken in September 2008 uncover how many Russians see the West as the primary instigator and the ones to blame for the short and successful Russian intervention. One poll shows that 32 percent of Russians blame the Georgian government, while 49 percent blame the United States and its desire to increase its

Table 3.9 Russian opinion of threatened NATO airstrikes on Bosnian Serbs

Respondents asked: Do you favor or oppose the proposed NATO airstrikes on Bosnian Serbs, and if so, would you consider it an attack on Russia, as the Serbs are fellow Slavic people?

	Yes	No	No opinion
NATO airstrikes favorable?	4%	68	19
Attack on Russia?	77	21	11

Source of data: Shiraev, Eric and Deone Terrio. 2003. "Russian Decision Making Regarding Bosnia: Indifferent Public and Feuding Elites." In *International Public Opinion and the Bosnian Crisis* (February 1994), Richard Sobel and Eric Siraev, eds. (Lexington: Lexington Books).

influence in an exclusive zone of Russian influence.[118] When asked why they think the West supported the Georgian side and not the Russians in the conflict, two-thirds of Russians believe that the West's intentions are to force Russia out of the Caucasus region completely.[119] Russian public mistrust of the West is confirmed by these polls, giving Russia the confidence to assert its authority in the former USSR, regardless of American alignment.[120] The United States was sent a clear and assertive message by Russia in 2008, and prospects for Georgian ascension to NATO are now bleaker than ever.[121]

Moving back in time to the immediate years after the fall of the Soviet empire, Table 3.10 gives the opinions of Russians about the Bosnian civil war of 1992–95. The multi-ethnic communist state of Yugoslavia was crumbling much like the Soviet Union, yet this transition to post-communism ended up becoming violent.[122] Ethnic groups that lived in relative harmony under the Yugoslav state and their latent animosities began to violently oppose each other for dominance and autonomy in many of the newly independent republics. Arguably the most violent outbreak of ethnic violence happened in Bosnia-Herzegovina, where Orthodox Serbs, Catholic Croatians, and Muslims took up arms against each other for control of the state.[123] Allegations of genocide and ethnic cleansing drew worldwide eyes, and the major powers of the world began to weigh in on the growing conflict. Western powers blamed the Bosnian Serbs for the escalating violence, making demands that they cease operations immediately or face NATO airstrikes.

Russia, which sees itself as the protector of the Slavic ethnicity and thus its fellow Serbs, maneuvered to protect its ethnic and religious brethren.[124] Public opinion polls taken right before NATO intervention in the crisis shows how strong Slavic ethno-nationalism permeates in the hearts and minds of most Russians. Seventy-seven percent of Russians opposed the proposed airstrikes, with only 4 percent approving. Furthermore, 68 percent of Russians saw the possible airstrikes as an

Table 3.10 Russian opinion of NATO intentions in the Kosovo crisis of 1999

To prevent atrocities on Kosovar Albanians, as NATO claims	14%
NATO has strategic military objectives to dominate the Balkan region	56
No opinion	30

Source of data: McLaren, Bronwyn. 1999. "Russia's reaction to Kosovo." *Global Journalist* (October). Accessed 8/12/2012, available at: http://www.globaljournalist.org/stories/1999/10/01/russias-reaction-to- kosovo/

attack on Russia and as an attack on a fellow Slavic people. However, Yeltsin had limited choices on what to do about these actions by the West, as his military institutions were bankrupt and in shambles at the time.[125] Five years later, however, when the state was better organized and the military capable to act, the Serbians once again needed Russian support in Kosovo.[126]

The Kosovo crisis in 1999 was another former Yugoslavia civil conflict with ethno-religious disputes that led to widespread violence. Kosovo is an autonomous part of Serbia that is ethnically different from the rest of the country; the majority of people there are of Albanian ethnicity.[127] Kosovar Albanians were demanding more autonomy from Belgrade to the point that an independence movement was becoming more and more popular. Kosovo is a place of great historical significance to the Serbian people, and this talk of autonomy and independence was met with the occupation of Serbian military forces to calm the rebellious fervor. Violence between the Albanians and Serbs ensued, and once again allegations of killings of innocents reverberated throughout the world.[128] The West took swift action, and NATO began bombing the Serb capital of Belgrade as well as other military targets in Serbia. NATO leadership demanded the withdrawal of all Serbian troops from Kosovo for the bombings to stop.[129] This time, however, Russian opposition to NATO actions was fierce. Table 3.11 shows that Yeltsin had public support to take such dangerous actions.

Why was Yeltsin so willing to stand up to the more powerful NATO as well as US President Bill Clinton, who until then had very cooperative and cordial relations?[130] The answers may lie in the public opinion polls presented in Table 3.11. More than half of Russians at the time saw it as a Western strategic move to dominate the Balkans and assimilate it into the Western world at the expense of Russian interests. Only 14 percent of Russians agreed with the stated NATO objective that it was to prevent atrocities committed on Kosovar Albanian civilians. This widespread anti-Western support by the Russian public may be behind Russia's

Table 3.11 Russian opinion of NATO bombings of Serbia during the Kosovo crisis

Support NATO bombings	2%
Oppose NATO bombings	92
No opinion	6

Source of data: Waal, Tom de. 1999. "Russian Public Demands Action." *BBC News* (April 9). Accessed 8/12/2012, available at: http://news.bbc.co.uk/2/hi/europe/315582.stm.

brinksmanship-style stance against NATO.[131] The results of Table 3.12 are even more telling.

Table 3.12 shows that 92 percent of Russians opposed the NATO bombings aimed at forcing the Serbian army to abandon Kosovo territory to stop the violent skirmishes and killings due to each side's close proximity. With only 2 percent approving, this is a near consensus for the Russian public, which could explain Yeltsin's hardliner stance against the United States and NATO. Negotiations between Russian and American diplomats lasted for months, with Russia not budging on the NATO airstrikes being halted immediately before any future agreements were to be made by the Serbs.[132] The American stance was that the airstrikes would stop once Serbian troops had withdrawn completely from Kosovo territory. The negotiations stalemated and in the meantime more people died from the bombing and Serbian-Albanian clashes in Kosovo.[133] Finally, an agreement was made and NATO as well as Russian peacekeeping troops moved into Kosovo and hostilities ceased.[134] However, Western and Russian troops hadn't been so close to each other in an adversarial situation since the Berlin Wall was still standing; this situation could have easily created misunderstandings and a very real possibility of armed conflict between Russians and Americans.[135]

In 2003, after US-Russian relations had a brief period of policy agreements in light of the 11 September 2001 attacks on New York and Washington, Russian coercive diplomacy against the United States and vice versa commenced to their usual levels of intensity. The row was over the internationally unpopular US-led invasion of Iraq. Russia, along with American allies France and Germany, saw the war as unnecessary and an affront to Iraqi sovereignty.[136] The condemnation of the US and United Kingdom increased as it became apparent that Saddam Hussein was not in possession of any weapons of mass destruction (WMDs), the major

Table 3.12 Russian opinion of the NATO intervention in Libya

	NATO intervention in Libya	Obama's handling of Libya
Support	20%	13
Oppose	64	54
No opinion	16	33

Source of data: Pew Global Attitudes Project Key Indicators Database. (Spring 2011). http://www.pewglobal.org/database/?indicator=27&survey=14&response=Unfavorable&mode=chart.
Russian Public Opinion Research Center. (March 2011). http://wciom.com/index.php?id=61&uid=44

argument through which the Bush and Blair Administrations justified the invasion.[137] Backed by several Western states, the Putin Administration opposed the Iraqi War.[138] This boisterous opposition was backed by Russian public opinion.

The Libyan civil war of 2011 began as a boost in confidence to opponents of the Gaddafi regime as a result of the successful and relatively peaceful overthrows in Tunisia and Egypt.[139] However, Gaddafi was not going to relinquish power without a fight. High amounts of reported violence and atrocities being committed by soldiers loyal to Gaddafi led to the involvement of international actors.[140] A unanimous UN Security Council resolution (China and Russia abstained) called for an end to the violence.[141] The UK, France, and Italy took the lead in a NATO intervention that bombed Gaddafi regime strongholds and armed the Libyan opposition with desperately needed weapons to turn the tide of the war in the rebellion's favor.[142] The United States took a backseat in this intervention and only contributed logistical and material support.[143] Russians believed that they had been duped, as their affirmative vote to stop the violence was not intended to increase the fighting and to give the go-ahead for NATO intervention. Russian and Chinese opposition to Western intervention became louder and more coercive.[144]

Table 3.13 shows the results of two polls taken during the conflict in the spring of 2011. The first asks Russians their position on the NATO intervention in Libya. Sixty-four percent of Russians oppose it; however, a sizeable 20 percent of Russians did support the end of the Gaddafi regime and backed NATO. Support wavered over time as NATO got more deeply involved, and eventually Gaddafi was killed and his regime toppled. Libya has been a close ally of Russia and the Soviet Union as a regime that stood up to the United States.[145] Similarly, the second poll asks Russians how they feel about President Obama's handling of the NATO intervention. Over half of Russians oppose the US president's handling of the situation, yet one-third of Russians are indifferent to the leader of Russia's principal rival and his Libyan policy. It seems that Russians did not care enough about Libya to put up a bigger fight with the West.

The final intervention analyzed in this section is still going on and has not reached a cessation of hostilities. We look at Russian opinion of the growing violence in the Syrian civil war that began with popular uprisings in the spring of 2011. Syria is one of Russia's best customers when it comes to the purchasing of Russian arms, military equipment, and technology.[146] Therefore, this time, when the UN Security Council came to vote on the cessation of hostilities and the reining in of the Assad regime, Russia, along with China, vetoed the Syrian resolution.[147]

Table 3.13 Russian opinion of the cause of the Syrian conflict

Respondents asked: What do you think is the primary cause of the situation in Syria?

Public dissatisfaction with Bashar al-Assad's regime	19%
Provocation by outside players such as the United States	48
No opinion	33

Source of data: Russian Public Opinion Research Center. (June 2012). http://wciom.com/index.php?id=61&uid=44

Western military action has yet to be taken, with only economic sanctions being the primary weapon against the Assad regime. Even so, Table 3.13 shows a June 2012 Russian public opinion poll about the cause of hostilities in Syria, with nearly half of Russians believing it to be Western provocation to get rid of a Russian friend. One-third of Russians are indifferent to the cause, yet this number could change if the militaries of the West get involved.

Another poll taken by Levada Center in June 2012 reports that only 29 percent of Russians believe their government should support the Assad regime in Syria and oppose regime change, which is the Western position.[148] These opinions may explain why Russia, although backing nonintervention by outside forces, has yet to take a harder stance against the West on the issue of Syria. Russia has stopped new sales of arms to the Assad regime as of August 2012 and has not mobilized any of its forces that are at its naval base on Syrian territory.[149] They are in support of UN envoys and negotiations with the rebels and the Assad regime.[150] Whatever the outcome of the Syrian civil war, it seems that Russia and the West will not escalate tensions over the Middle Eastern country and use it as a focal point for competition over global leadership. Only when their fellow Slavs, the Serbs, were the target of Western military intervention, it seems, were Russians willing to go to the brink of armed conflict with the West. Ethno-nationalism is an important part of the Russian psyche, which is an important finding of this chapter.[151] When the fate of a Slavic-Orthodox people is at stake, Russia will come calling, and with force. However, there is one more important policy area that the Russian public finds salient and that may explain their peculiar foreign policies with their number-one export: energy.

Conclusion and Future Termination Paths

Russia views itself as a great power.[152] Whereas in the 1990s it was relatively weak, Russia has since rebounded and returned to following the dictates of its traditional saying, "Russia has no friends except for its army and navy."[153] In response to a world order not consistent with the hopes for the post-Cold War era, Russia has reverted to the old doctrines of nationalism and military power. However, as discussed in the introduction of Chapter 1, Russia does not have the capabilities and demography to have the global reach it once had and to essentially "put its money where its mouth is." However, the country has embraced its newfound economic leverage (to be explored in Chapters 5 and 6) and multilateralism through which to contain the ambitions of the United States in post-Soviet regions. These developments revived the rivalry pronounced dead in 1991.

The United States has seemed to wake up to the threat of Russia after 1994 and has pushed democratic reforms, expansion of alliances, and arms transfers and training in Moscow's perceived exclusive sphere of influence. The rivalry's persistence despite regime change and the end of the Cold War demonstrates the condition of addiction as necessary for evidence of rivalry. Further, the relative positioning arguments regarding post-Soviet space provide evidence that the rivalry dominates strategic concerns and also that it has shifted to a policy rivalry focus. Yet neither side remains blameless for the continuation of the rivalry. While the rivalry went on the backburner in the period immediately following the Cold War, it has continued to survive throughout the 1990s (e.g., Bosnia, Kosovo) only to be reignited again due to developments in Russia's Near Abroad (e.g., NATO enlargement, Color Revolutions). It may not be the level witnessed during the Cold War, but the current period is nonetheless dangerous. Scholars and policymakers have to be aware of the impact of actions in the context of rivalry. The continuation of this rivalry has immense implications that seem to be ignored in the Age of Terror.

How can this rivalry end? Vasquez points to four paths for peace in the post-Cold War era. Both Russia and the US need to establish the rules of the game (of global politics), end reliance on power politics (i.e., arms races and alliance systems), learn how to better deal with territorial disputes, and embrace new practices to settle disputes.[154] This analysis shows that the Vasquez criteria for peace in a rivalry have not been met. First, the rules of the game have fluctuated widely since the end of the initial cooperation period in 1993. Second, the two systems still use power politics. The United States continues to rely on NATO and other forms of threat to change Russian behavior. Conversely, Russia has cultivated a closer relationship with China and the East broadly conceived through its involvement in the Shanghai Cooperation Organization.

Russia has also increased its military budget steadily after initial economic setbacks.[155] Third, Russia continues to assert territorial claims in the former Soviet republics. For example, Moscow provides active support to separatists (through money, training, and even passports) in Eastern Ukraine, Abkhazia, South Ossetia, and Transnistria. Fourth, there have been no new practices for the settlement of disputes. Although the US and Russia cooperate more on economic issues, the two countries often differ on security topics and how to settle these disputes. Perhaps the most obvious example is the 2003 Iraq invasion. Washington and Moscow took diametrically opposed positions on the war. In the end, the disagreement was "settled" due to *fait accompli* rather than negotiations between the two powers. The end of the rivalry looked likely in 1991, and to this day, many still view and code the US-Russia rivalry as terminated. However, domestic political variables have prevented the establishment of new norms that Vasquez calls for to maintain peace and end the rivalry.

Through the examination of domestic conditions and foreign policy concerns, we argue that the rivalry has continued into the post-Cold War world in regional theaters over territorial and policy issues. People change slowly. So do rivals. As the rivalry continues, it remains dangerous. The initial stage of the rivalry, from 1947 to 1991, produced no direct confrontation. It still gave rise to conflicts on the periphery and raised the possibility of a nuclear confrontation throughout its duration. The recent stage of the rivalry, which began in 1993, has also produced no direct confrontation to date. Yet, the main danger is that it will prevent democratization in Eastern Europe and continue to facilitate ethnic conflicts as the US and Russia compete for influence in post-Soviet space. This last point became only too clear with the Russian intervention into Georgia over the South Ossetia region as well as its involvement in the destabilization of Ukraine. NATO involvement in Slavic affairs has also been a point of contention between the US and Russia.

The recent adversarial take on Ukraine, Libya, and Syria will continue to lead to problematic relations between the US and Russia as they compete to express their points of view. Stands on these conflicts have led to tit-for-tat domestic policies directed at one another as a result of this escalating rivalry. Recent US legislative acts directed at alienating Russia include the 2012 congressional passage of the Magnitsky Act, which bans all Russians suspected of human rights violations from entering the US, as well as the disallowance of the Russian version of the GPS system setting up towers in the United States so as to become more globally competitive.[156] This is a stark and illustrative example of their continued adversarial behavior. Russia has countered by passing a law that bans the adoption of Russian orphans by American citizens, as well as harboring the well-known former NSA contractor Edward Snowden.[157] The continued rivalry decreases the chances for conflict resolution in these areas and only makes the resumption of disputes over the "frozen conflicts" all the more likely. However, the regional and international issues are only half the story in explaining the continued US-Russian rivalry. Rivalry plays an integral part in this empirical theory on armed conflict. One cannot understand Russian foreign policy without understanding the dynamics of the ongoing US-Russian rivalry. The next four chapters will use the factors of situational coercive diplomacy presented in this and the preceding chapters and examine three important issues regarding Russian foreign policy tactics: cyber conflict, energy coercion, and Russia's involvement in the fate of the Far North as maritime policy. How are these tactics used, and when does Russia employ them? How does the US-Russian rivalry fit in to these decisions made by the Russian foreign policy elite? How are issue salience, rivalry, public opinion, and great power identity important to these decision-making processes? We now move on to see if the same dynamics of Russian coercive diplomacy are used in another issue realm—cyberspace.

4
Russia in Cyberspace

Introduction

In April of 2007, the Estonian government removed a Soviet-era grave marker, "The Bronze Soldier of Tallinn," from the center square of the city to a more remote location. This statue marked the struggle of the Soviet soldiers against Nazi invaders during World War II. A symbol of Russian pride, the statue has less significance to ethnic Estonians, as their country was engulfed by the Soviet Union following the expulsion of the Germans.

The Russian response to the memorial's removal was swift; a flurry of complex, organized, and widespread malicious cyber operations flooded both private and public Estonian networks.[1] From April 27 to around May 10, Estonian commerce suffered setbacks, as banks and businesses were temporarily offline. ATMs could not be accessed, and retail outlets were forced into cash-only polices for a number of days. The Estonian government was mostly incapable of operating, as many government services, ranging from vote registering to licensing renewals, are exclusively conducted online in the Baltic state. Furthermore, Estonian citizens could not conduct their daily activities, such as e-mail exchanges or social networking, on the Internet.[2] This was recognized by some as the second most sophisticated cyber operation to date (after Stuxnet), and the outcry from Estonia and much of the West was widespread.[3]

The following year, in August of 2008, a series of cyber operations was launched against Georgia, running simultaneous to the physical military intervention Russia launched against the former Soviet republic. Georgian government websites were vandalized with pictures of President Mikheil Saakishvelli appearing as Hitler;[4] other sites were flooded with DDoS methods, or denial of service tactics that overload websites with too many users, and shut down.[5] US government websites in Georgia were also hacked and hijacked with instructions to infiltrate the Georgian interfaces. The effects of these operations were widespread confusion among Georgian and American officials in Tbilisi, the Georgian capital, who could not decipher what was actually going on with regards to the movement of soldiers and military hardware.[6] These cyber incidents within the context of a physical military intervention were the first of their kind; no state had used cyber tactics when implementing a military campaign; even the United States refrained from doing so during the Libyan civil conflict in 2011 and the Iraqi invasion in 2003. It seems that the Russians were opening the feared "Pandora's box" of global cyber warfare.

These malicious operations in cyberspace orchestrated by the Russian government suggest that the Russians will not hesitate to use their cyber capabilities under certain contexts. However, Russia does not use its capabilities in cyberspace often, nor does it use them gregariously and to their full potential. Rather, we argue here that Russia uses its cyber technology against other states almost exclusively in post-Soviet space, against states it already has a rivalry with, and it is constrained from using cyber tactics elsewhere by the fear of retaliation. Even when it uses these tactics, it hesitates from extreme action. Cyber conflict is the least costly tool for Russian foreign policy. Since Russia could claim it did not launch these operations due to the attribution problem in cyberspace, cyber tactics seem to be the simplest choice available when the coercive path is taken.

Russia's cyber malice is quite tame when compared to its conventional potential and threats in the region. Russia has been restrained to some extent, and Estonia has responded not with fear and further threats but by becoming the international leader in developing positive cyber norms and shaming the Russians.

This chapter will discuss how cyber conflict is conceived and analyzed in popular and academic discourse, followed by a section uncovering who is behind Russian cyber capabilities. Then we will show how the capabilities Russia contains compare to other cyber "heavyweight" states that have advanced technological capabilities, followed by an analysis of how Russia has used this technology as a foreign policy tool against its Near Abroad adversaries of Estonia and Georgia, as well as its long-time rival, the United States.[7] A discussion of the use of cyber tactics by Russia in the 2014 Ukrainian crisis will be included in the concluding section.

As the face of diplomacy changes with the advent of cyber capabilities, it is important to understand how this new tool is used and when, if at all. Cyber strategies are part of the arsenal of states' coercive capabilities, but as we will document in this chapter, Russia tends to utilize these tactics only when the issues are salient to the state, and Russia fails to maximize its cyber capabilities.

Russia and Cyber Conflict

Several definitions of cyber conflict abound, such as "hostile actions in cyberspace that have effects that amplify or are equivalent to major kinetic violence"[8] and "penetration of foreign networks for the purpose of disrupting or dismantling those networks, and making them inoperable."[9] We define cyber conflict as the use of computational technologies in military interactions or diplomatic affairs in the realm of the international system.[10] The term "cyberwar" is not meant as a war in the traditional sense but as a tool to use in foreign policy. The term has taken on a life of its own, and despite its inaccurate nature, we continue to use it due to its commonality in language. In fact, there have been no cyber conflict deaths as of yet, and it is unclear if there ever will be.[11] Instead, our usage of the popular term focuses on its ability to alter perceptions and intentions of states. Like any foreign policy tool, its purpose is to alter behavior.

Russia uses cyber conflict along similar lines, as a coercive tool. Those on the Russian General Staff define cyber conflict as "disruption of the key enemy military, industrial and administrative facilities and systems, as well as bringing information-psychological pressure to bear on the adversary's military-political leadership, troops and population, something to be achieved primarily through the use of state-of-the-art information technologies and assets."[12] For Russia, cyber conflict is a power politics strategy to be used against an enemy to achieve desired ends.

Cyber conflict is not limited to state-to-state interactions, which is the focus of this chapter, but can come from individuals, terrorist organizations, corporations, and other non-state actors. Joseph Nye delimits three domains of cyberspace: 1) governments, 2) organizations, and 3) individuals.[13] Government cyber conflict is directed by government actors and foreign policy decision makers with the purpose of disrupting the functionality of other governments. Organizational cyber conflict typically involves organized non-state actors such as "hacktivist" groups with a political agenda, like Anonymous. Individual-based cyber conflict would cover rogue actions by lone operators either online to steal information and sell it to the highest bidder or to cause minor chaos in a network.

"Cyber security" is the term used for a state's defensive capabilities in cyberspace. Libicki defines cyberspace with four requirements that ensure that the concept is "1) replicable, 2) consists of recognized actions, such as text in English, codes understood to humans as opposed to binary code which most cannot understand, 3) tends to have persistent rules or technologies, and finally, 4) is divided between the physical layer and syntactic layer that is information and knowledge."[14] It is how well a country can defend itself from a cyber incident. For example, China has the "kill switch," or the ability to shut off all international Internet lines going in and out of its borders at any time.[15] Cyberspace in the United States, on the other hand, is controlled by many private parties, and defense in the face of a cyber incident would be nearly impossible to coordinate. Russia has an online infrastructure where most Internet traffic coming in and out of the country goes through the networks of the Federal Security Service (Federal'naya sluzhba bezopasnosti) or FSB. The owners of Russian networks and security apparatuses are also part of the close and elite circles within the Putin Administration. Russia's offensive capabilities are also impressive: It was a Russian firm named Kapersky Labs that helped find the now infamous Stuxnet and Flame malware incidents in Iran. It can also be considered that offense is the best defense and that cyber security also relates to the use of offensive capabilities to deter future cyber incidents.

At this point, it is important to outline the different types of cyber methods and the amount of damage they can potentially inflict on international actors to facilitate analysis. In a previous article, we construct a typology and severity scale for cyber incidents and disputes among contemporary rivals.[16] Our severity scale ranges from one, which indicates minimal damage, to five, which indicates state catastrophe as a direct result of a cyber operation. Vandalism or website defacements are the first method of cyber tactics.[17] Potential hackers use command codes, or language that instructs computers to perform certain actions, known to computer experts to deface or destroy a target's web page(s).[18] Although rather benign, this method may have important psychological effects. Vandalism is usually given a severity score of one.

The second method for hackers is dubbed the distributed denial of service (or DDoS). This technique is not sophisticated and can be used by the most novice hackers with malicious intent. At this point there are now websites that help make the process easier and automated. Denial of service methods flood particular networks with more requests for data than the site can process.[19] The effect of such a tactic is to effectively shut down the site, thus preventing both access and usage. Important government sites can be shut down and give a country much grief for a number of days. Such methods are coordinated and implemented through "zombies" or "botnets," a network of computers that have been forced to operate on the commands of an unauthorized remote user, in other words, the coordinated use of computers by like-minded hackers working in unison.[20] Disruption of service and general widespread malice are the primary effects of these methods. These methods are usually given severity scores of only two.

Intrusions are the third option for cyber warriors. Trapdoors, trojans, or backdoors are unauthorized software added to a program to allow entry into a target's network for future access to a site.[21] This can be more damaging than DDoS methods or website defacements, as the intrusions can be numerous and escalating before the targeted network's operators even realize that they have been compromised. The purpose of trapdoors is to steal sensitive information from secured sites. This method usually warrants a score of three in the severity range.

Infiltrations are the most potent method available to hackers, and the damage inflicted can be widespread and potentially catastrophic. There are several types of infiltrations. Examples include logic bombs, worms, viruses, packet sniffers, and advanced persistent threats (APTs).[22] These methods force computers or networks to undertake tasks that they would normally not undertake. Infiltrations are usually the most malicious in intent and have the capability to launch missiles, shut down infrastructure, and have catastrophic effects on a state's national security. These methods can range anywhere from two to five on the severity scale; however, up to this point, the highest score recorded has been three.[23]

With these concepts and categories in mind, we now move forward to uncover how Russia uses its capabilities in cyberspace. Russia is a formidable force on the cyber battlefield, yet it seems to use its arsenal exclusively in its physical sphere of influence of the former Soviet Union. The next section examines who is behind Russia's cyber offensive and defensive capabilities.

Russia's Cyber Capabilities

Along with the powerful private firm of Kaspersky Labs, the FSB security service in Russia closely regulates the Internet within Russia's borders.[24] Many contemporary Russian defense experts note that the Soviet Union did not give cyber capabilities much thought as it rose in importance in the 1980s, and, without a hint of satire, argue that this is a contributing factor to the former empire's demise.[25] The Putin Administration does not want to make the same mistake and has given the FSB the funding to monitor the Internet in Russia as well as allow it to develop Russian offensive capabilities. To quote the Doctrine of Information Security of Russia, "The national security of the Russian Federation depends to a substantial degree on ensuring information security, a dependence that will increase with technological progress."[26]

The FSB's role in Russian cyberspace is widespread and might be considered extreme compared to its Western counterparts. All Russian Internet service providers (ISPs) are required to install hardware to allow the FSB to monitor the Internet usage and e-mail messages of all its customers.[27] All data used by these ISPs are also required to go through FSB computers to be recorded and archived. Furthermore, foreign companies such as Microsoft are required to share their source codes with the FSB.[28] In the name of national security, therefore, the FSB is monitoring all users of the Internet within Russian borders, a tactic prohibited by US law unless there is a link to terrorism or organized crime.

Monitoring the Internet is a daunting and near impossible task. Therefore, the Kremlin instituted a second tier of protection and support by private firms. One firm is the cyber security giant Kapersky Labs. In May 2012, this Internet security firm, the Russian equivalent to Western firms such as McAfee and Symantec, uncovered a new virus that has ties to the famous Stuxnet worm, the cyber intrusion noted for temporarily crippling Iran's nuclear program.[29] Dubbed the "Flame" virus, Kaspersky found that it is 20 times more sophisticated than Stuxnet. Kaspersky believes that the virus, tied to the Stuxnet worm, has been in operation since March of 2010. The Russian firm argued that the likely culprits behind the virus were the United States and Israel, two of the most vocal opponents of Iran's nuclear program.[30] Publicizing the intrusions and infiltrations was a national security tactic to demonstrate that the West is willing to use cyber tactics as a foreign policy tool while Russia is a more moderate cyber power. Flame appears to be an early test of possible cyber capabilities and not the deadly virus the Russians made it out to be. Its capabilities seem relatively trivial unless connected to a much more powerful set of operations. Nonetheless, the media, in both the West and in Russia, consumed and parroted the information with little analysis.

Kaspersky Labs' uncovering of this virus shows how private Russian Internet security firms can be a foreign policy tool. The firm has over 300 million customers worldwide and has sales equivalent to those of McAfee and Symantec combined. It is by far the world's largest Internet security firm.[31] It has publicized several of the United States' cyber capabilities. Kaspersky Labs operates in tandem with and for the interests of the Russian government in cyberspace.[32] With the backing and support of the Russian government, the firm can enhance the power of Russia in cyberspace. It is also an important counter to and restraint for Western cyber power.

Kaspersky Labs, unlike its Western counterparts thus far, is capable of discovering and perhaps foiling Western cyber espionage. What is more troubling to Western cyber powers is that Eugene Kaspersky is a former member of the KGB, a loyal supporter of Vladimir Putin and his government, and has a close relationship with Russia's FSB, the successor to the Soviet-era KGB. Furthermore, Kaspersky promotes security over Internet freedom, and his company has been behind the thwarting and censoring of social media networks critical of Putin.[33]

Along with Kaspersky, many other pro-government conglomerates have bought up the digital infrastructure in Russia to ensure its security and ease of monitoring. The most popular social networking site in Russia, LiveJournal, is owned by government-friendly business interests.[34] State-controlled energy giant Gazprom's media subsidiary, Gazprom Media, has been buying up many Internet companies, as have many of the rich and elite oligarchs loyal to Putin. These companies and individuals have a seamlessly friendly relationship with the FSB, which indicates that elite private citizens are willing to cede their business interests in the name of Russian national security, or perhaps in the name of personal security.[35] Russia, therefore, has a formidable infrastructure in cyberspace that is closely controlled and monitored either by the state or companies loyal to the state. Any deviation from this path is a sure method of corporate suicide.

Russia also has a disproportionate amount of "hacktivists," or politically active and technologically savvy individuals with a high degree of Russian patriotism.[36] It is speculated that hacktivist groups, allowed by the tightly controlled Internet-monitoring FSB, were behind the flood of DDoS incidents on Estonia in 2007.[37] An army of cyber patriots whose government turns a blind eye to this kind of cyber mischief can enhance a state's cyber power if a true cyberwar were ever to escalate.

Russia's official government statements about the cyber operations against Estonia in 2007 and Georgia in 2008 was that it was not directly involved, and that Russia has an army of these cyber patriots or cyber nationalists who were standing up for Russian dignity when their homeland was insulted or threatened.[38] Indeed, this was hard to dispute, as tracing the origins of these hackers was found to be difficult. As Stuxnet and Flame's origins were unable to be deciphered, it is still a mystery as to who exactly was behind the incidents on Estonia and Georgia. This makes Russia a force in global cyberspace and indicates that Russia is ready to fight the inevitable "Cyber World War" many media pundits and security experts in the West say is on the horizon. The next section analyzes how Russia stacks up against the other cyber powers of the world and finds that it may be the most "dangerous" state in the world in terms of cyber capabilities.

Russia's Place in the Realm of Cyber Powers

Russia is considered one of the "heavyweights" in advanced cyber capabilities.[39] Its contemporary domestic initiatives and its perpetual competition with the United States have left the new Russian state technologically savvy and endowed with a population of computer experts. Even though Russia is not as technologically advanced as the United States, when it comes to cyber capabilities, Russia is a force with which to be reckoned.[40] There are only a few cyber heavyweights recognized by cyber security experts of the security, academic, and media realms.[41] Among these are Russia, the United States, China, Iran, North and South Korea, the European powers of Great Britain and Germany, and Israel.[42] Table 4.1 shows the capabilities of these states and finds that Russia is the most "dangerous" state in cyberspace.

Table 4.1 is an outline of cyber capabilities as adapted from Clarke and Knake's scale in their 2010 book and combined and updated with Booz Allen Hamilton's Cyberhub Capability Index.[43] It denotes how offensively capable, dependent upon cyberspace, and defensively capable each country is. Each value is based on a scale of ten. Offensive capabilities measure how technologically advanced a state is in using the Internet as well as the endowment of technologically trained citizens and workers it has. Cyber defense is scored based on how much control a state's government has on regulating the digital information coming in and going out of the country. It also takes into consideration whether or not the state has the "kill-switch" capability where all incoming data from international cyberspace can be shut down at a moment's notice. Cyber dependence is ranked in reverse order, as the higher the number, the less dependent the state is on computer networks. The United States gets a low score because of how "plugged in" it is to the web for important infrastructural needs, such as electricity and water.[44] Therefore, the more dependent a state is on cyber technology, the more vulnerable it is.[45]

Iran is considered a heavyweight, but only in reaction to the cyber incidents that it receives from supposed Israeli and American infiltrations. Iran has assembled an elite squadron to defend itself from recurring incidents and to go on the offensive against Israel. These defenses

Table 4.1 Overall cyber capabilities among states

State	Cyber-Offense	Cyber-Dependence	Cyber-Defense	Total Score
Iran	4	5	3	12
Great Britain	7	2	4	13
Germany	7	2	4	13
South Korea	6	4	4	14
North Korea	3	9	2	14
United States	9	2	4	15
Israel	8	3	4	15
China	6	4	6	16
Russia	7	3	7	17

Source of data: Conceptualized and altered from Clarke, Richard A. and Robert K. Knake. *Cyber War: The Next Threat to National Security and What to Do About It.* 2010. (New York: HarperCollins, Inc.): 148; and "Booz Allen Hamilton's Cyberhub Power Scale." Accessed 8/2/2012, available at: http://www.cyberhub.com/.

have been shown to be weak, and Iran's offensive capabilities are merely to vandalize or enact minor DDoS tactics. The European powers of Great Britain and Germany are, like the United States, very "plugged in" when it comes to dependence on cyber technologies and are also very offensively capable. South Korea has expanded its cyber capabilities in reaction to the malicious operations from its longtime rival from the north. North Korea, which is technologically backwards in many respects, has begun a cyber offensive campaign to menace South Korea and the United States. Retaliation against North Korea in cyberspace has proven difficult, as most of its infrastructure and citizenry are not online. Israel is very capable offensively and has used these capabilities to set back Iran's nuclear program with help from the United States. Finally, China is the most active state in cyberspace, menacing its East Asian rivals, but it is mostly a thorn in the side of the most technologically advanced state in the cyber realm, the United States. The United States has had interactions in cyberspace with Russia; therefore, we compare the capabilities of the Cold War adversaries and current rivals.[46] The United States can be and is the state most frequently infiltrated by potentially malicious software. Furthermore, as the world's sole superpower, the United States can be and is the main target state that dissident groups, terrorists, and rogue states wish to damage. Almost all infrastructures are now connected to the Internet in the United States; therefore, the number of targets for malicious cyber operations is nearly unlimited. Thus the expectation is that the United States, with its dependence on the

Internet, would have advanced cyber defenses, with government playing a leading role in the protection of both private and public domains. However, this is not the case. Due to the American creed of privatization and free information, the American network remains quite vulnerable. In reaction to these vulnerabilities, the recent "kill-switch" bill in Congress proposed by many on both sides of the aisle failed to pass both in the Senate and House. Therefore, America's cyber defenses will remain vulnerable, yet it remains to be seen how threatening this vulnerability actually is.

Looking at Table 4.1, the United States gets a cyber dependence score of two and a defensive score of four. The United States, although potentially vulnerable, is also considered the most offensively capable cyber state in the world. Its offensive capabilities are ranked the highest in the world and are given a score of nine. It is speculated, and for the most part backed up by evidence, that the famous Stuxnet worm and Flame virus were developed in the United States and then given to Israel for deployment into Iran's nuclear network. These incidents are considered the most sophisticated cyber tactics to date, indicating that the United States may be well ahead of the rest of the world when it comes to offensive capabilities. This advantage offensively, it can be argued, makes up for American vulnerability on defense, as no state would want to feel the wrath of American cyber retaliation. The US has a vast team of government cyber operatives, and the entire population could be mined and mobilized to fight a cyber battle.

Russia gets the highest score on our cyber capability table, and therefore can be considered the most "dangerous" cyber power in the world. It has offensive capabilities that nearly match those of the United States and Israel, and its relatively low dependence on networks for critical infrastructure makes it hard to infiltrate from the outside. Along with FSB monitoring and a highly advanced security firm loyal to the government, Russia is ready to defend itself against a severe cyber infiltration. It therefore receives a high defensive score of seven, the highest defensive score of all the cyber powers. As we will see in this chapter, Russia also has the willingness to use cyber capabilities against international enemies.

Russia also has a highly educated and technically skilled workforce, which are remnants of the Soviet era and its huge defense industry. Furthermore, Russian culture demands a certain degree of nationalistic pride, as thoughts of the glory of the recent Soviet past still loom large in the hearts and minds of most Russians. High-tech jobs are not as plentiful in Russia as they were during Soviet times. This combination of few jobs in the Russian private economy for high-tech skills along with Russian national pride has created a large market of nationalistic hacking communities that have a potent ability to inflict damage on states. The capabilities of the offensive Russian cyber community is therefore massive, slightly reckless, and has the potential to do great damage to those offending the Russian state.

Russia has, therefore, the capabilities to start the often-feared cyber armageddon that many media pundits and industry experts say is inevitable. If cyber conflict is to be used as a means of power politics and an effective foreign policy tool, we should see plentiful Russian cyber operations in the former Soviet Union and globally. However, this is not the case. Russia has only used its cyber arsenal on Estonia and Georgia in the Near Abroad and very sparingly globally, on its longtime rival the United States (but in Afghanistan). The theory outlined in Chapter 2 suggests that coercive foreign policy tactics will only be used in two situations: when the issue is highly salient and when there is an ongoing rivalry. Both conditions are met in the following case studies.

Cyber Conflict in Post-Soviet Space and Beyond

Finding and categorizing cyber incidents into an organized dataset can be daunting. The best method for coding Russian cyber operations for this chapter is to search the archives of news stories in the Google News search engine and also to comb through reports, books, and testimonies of cyber disputes. However, attribution of cyber tactics can also be problematic. One of the advantages of cyber tactics is deniability. Russia is famous for denying responsibility even when all of the guilty signs point to Moscow. Russia appeared to have coordinated its infiltrations against Georgia, and it has not denied its part in the operations. It first denied the cyber operations on Estonia only to retract and admit responsibility once the situation had cooled down. Based on a wider dataset collected, we enter the search query "Russia" AND "victim state (e.g., United States)" AND "cyber" OR "Internet attack" OR "infrastructure attack" OR "government attack."[47] What we wish to uncover in this search is the date and duration of the incident, who initiated the incident, and the type and severity of the incident. The time period is from 1 January 2001 to 31 December 2011, so we could get a ten-year sample and also capture the main period of active and growing Russian Internet engagement. We find that only three states have been the victims of Russian cyber malice: Estonia, Georgia, and the United States. Table 4.2 shows the summary of our findings for cyber incidents involving Russia.

Table 4.2 Summary of cyber incidents involving Russia 2001–11

Dyad (Initiator First)	Name (Duration)	Type	Severity Score	Explanation
Russia–Estonia	"Bronze Soldier Retaliation" (4/27/2007–5/10/2007)	Vandalism, DDoS	2	Response to Estonian removal of a Soviet-era war memorial, widespread DDoS and vandalism
Russia–Georgia	"Before the Gunfire" (4/20/2008–8/16/2008)	DDoS	1	Ongoing DDoS tactics before the Russo–Georgian conflict
Georgia–Russia	"Osinform.ru Website" (8/4/2008–8/4/2008)	DDoS	1	Russian "hacktivist" networks shut down after Georgian troops killed
Russia–Georgia	"VoiP Phone System" 8/4/2008–8/8/2008)	DDoS	1	Infiltration of major Georgian mobile network
Russia–Georgia	"Georgian Government Site Defacements" (8/7/2008–8/16/2008)	Vandalism	1	Widespread vandalism on Georgian government sites before conflict
Russia–US	"US Identities Stolen to Hack Georgia" (8/6/2008–8/12/2008)	Infiltration	1	US government sites hacked to steal identities and vandalize Georgian sites
Russia–US	"US Central Command in Iraq and Afghanistan Hacked" (11/26/2008–11/28/2008)	Infiltration	2	Information stolen from US Central Command, origins in Russia
Russia–US	"US Power Grid Hack" (8/24/2009–Ongoing)	Infiltration	3	Eastern Seaboard power grid hacked but no damage, origins in Russia
Russia–US	"Dragonfly Energy Grid Hack" (1/15/2013–Ongoing)	Infiltration	3	Hacker group originating from Russia infects the US energy grid

Estonia

After nearly 50 years as a Soviet republic, Estonia became an independent state in 1991. Estonia today is considered to be hostile to Russia. Perceptions of the history of Russian mastery over Estonia are at odds on each side of the border.[48] These perceptions date back to the end of the Second World War. Russia sees itself as a liberator of the Estonian people from Nazi rule and oppression. Estonians perceive these events differently; they interpret Russia's engulfing of their state as replacing one despotic ruler, Stalin, for another, Hitler.[49] These different takes on the fate of Estonia came to a head with the removal of "The Bronze Soldier," a Soviet war memorial honoring the dead of the Red Army who expelled Hitler's forces in 1944. Much of the Russian population and government found this removal offensive, as they perceived this action as dishonoring those who gave their lives fighting the Nazis.[50] Estonians saw the removal as part of the process of moving on from their Soviet past. On 27 April 2007, Estonian Internet networks were flooded with vandalism and denial of service operations originating from Russia. It is estimated that losses of around $750 million resulted from these operations.[51]

Estonian government and private networks were effectively shut down for about two weeks. Estonians could not use bank cash machines, the government had trouble conducting business on the Internet, and commerce in the country suffered through a lack of Internet connectivity (credit card machines are often connected to banking networks). We give this brief and not very disruptive cyber operation a severity score of two. Although the incidents were widespread, they did not target a specific state strategy of Estonia; they only caused widespread yet relatively undamaging confusion and difficulty. The cyber operations were upsetting to Estonia and unsettling to the West. Many sites were flooded with Russian propaganda and false apologies. Others were DDoS methods that flooded websites and effectively shut them down. Estonia is known as the most "plugged in" European country, as the most businesses, government entities, and citizens per capita are dependent upon the Internet. These cyber operations, therefore, affected most of the population, yet citizens could also combat most of the problems by operating as many did a few short years ago (based on cash and barter systems). Although these tactics were widespread, the long-term damage from these operations is, for the most part, nonexistent.

International condemnation of the 2007 operations was widespread. The United States, EU, and other Western powers scorned Russia for its abuse in cyberspace.[52] The US House of Representatives went as far as to pass a resolution condemning the violent protests by ethnic Russians in Estonia as well as Russia's use of its cyber power against an otherwise peaceful country that respects the rule of law.[53] The more important outcome from these incidents is not that Russia was scorned for these actions but rather the surprising behavior from the victimized state. Estonia did not retaliate against Russia in cyberspace. It did not seek help from its powerful NATO or EU partners to stand up to the Russians. It did not request economic sanctions, military sabre rattling, or help from the United States in constructing a destructive cyber counterattack to show Russia that it cannot be targeted. Instead, Estonia sought a world forum to discuss its case of victimization in cyberspace. Since the 2007 incidents, the Estonian capital of Tallinn has hosted the "International Conference on Cyber Conflict" four times. It is an outcropping of NATO and hosts the major powers of the West as well as many minor powers in Europe.[54] These conferences have promoted the adoption of cyber norms and modes of behavior in cyberspace so that conflicts do not escalate as a result of malicious operations from the Internet. These include extending the notion of territorial sovereignty to states' cyberspace; for example, Article 51 of the UN Charter allows proper action to be taken for defence against cyber tactics in lieu of an armed attack and the categorization of cyber incidents according to low or high intensity so that the proper international responses can be taken.[55] Estonia, a country with a population of just over one million, has become a global leader in the promotion of liberal and democratic practices and norms on the Internet. It has informed much larger states that "getting even" is not the way to deter future incidents; instead, setting up rules and norms in cyberspace is the most constructive way to keep states, democratic and autocratic alike, in line.[56]

This form of deterrence in cyberspace has been a centerpiece of Estonian foreign policy and how Estonia is being recognized by the world. The actions taken by Estonia have kept Russian hackers at bay, as it has not suffered a Moscow-led cyber operation since (nor has it done much to provoke one, to be fair). If another course of action was taken, perhaps Estonia could have felt the full force of the Russian cyber arsenal. However, Estonia's behavior after it suffered these operations can only be admired, and this measured approach by a victimized state should be the desired behavior of most states that suffer from cyber malice. The condemnation by the international community could deter future incidents and prevent the widespread abuse of cyber power.

From the Russian standpoint, Moscow can be praised for showing restraint as well. Based on our cyber capabilities rankings in Table 4.1,

Russia could have done a lot worse to Estonian networks. It could have injected a sophisticated worm or virus along the lines of Stuxnet or Flame, the world-renowned forms of malware that have set back Iran's nuclear program. It could have knocked out Estonia's power grid or water supply, both hooked up to the Internet and easily penetrable by Russian hackers. It could have cut off gas supplies if the row happened during the cold winter months. For all the international backlash Russia received for these operations, it did demonstrate some restraint. Why?

We assert that although Russia was offended by Estonia's removal of a symbol of Russian pride and perhaps wanted to punish Estonia more severely, Moscow was restrained from taking further steps in restoring its honor and pride. Estonia had attained full EU and NATO membership by 2007. It would probably have been unwise of Russia to implement more severe cyber tactics, energy sanctions, or military mobilization in response to Estonia's actions. A more severe cyber operation at the time would have crossed Russia into unknown foreign policy territory. Russia would have been the first government to release a malicious cyber operation such as a power grid knock-out or a sophisticated worm such as Stuxnet, which happened after these events. This event could have opened the proverbial and literal Pandora's box that may have escalated the dynamics of cyber conflict globally. Starting this form of escalation and new way of warfare against a member of the EU and NATO would probably have been unwise.

Another reason that Estonia probably did not retaliate is because Russia, as shown in the next two chapters, has been quite effective in using its energy power as a foreign policy weapon. Estonia is completely dependent on Russian imports for its natural gas supply. Luckily for Tallinn, the Bronze Statue row happened during the spring, a time when the Estonian climate is mild and the effects of a natural gas shutoff would have also been quite mild. Had this dispute happened during the winter months, Russia could have used its coercive energy tactics and shut off gas supplies to Estonia, as occurred in Ukraine and Moldova. Russia also supplies oil to Estonia, and it did halt shipments to Estonia for a few days during the time of the cyber incidents. However, Estonia began courting other oil suppliers such as Norway to increase supply, and Russia quickly resumed exporting oil to the Baltic state lest it lose a close and easy customer. Thus economic statecraft was not a viable option for Moscow.

Threats to use military force over the war memorial dispute would have been considered a highly disproportionate response and could have provoked a NATO response. Russia, although highly offended, would probably not want to draw the United States into the row over the movement of a statue. Estonia, for its part, did not invoke its NATO membership status over the cyber operations and instead shamed Russia publicly and globally for its behavior. Instead of an Estonian cyber retaliation that would have infiltrated Russian networks and escalated the dispute, Tallinn refrained and became a leader in promoting norms and accepted modes of behavior for state-to-state cyber interactions. Although Russia and Estonia have yet to completely reconcile over the dispute, escalation of the row has been averted, and Russia has been deterred from continuing cyber operations against its former vassal state. The following year, in 2008, however, Russia's cyber arsenal struck again, this time in tandem with its military attack on the ex-Soviet state of Georgia.

Georgia

Details of the buildup to the conventional military Russo-Georgian conflict of 2008 are covered in the previous chapter. Chapter 3 also shows that Russian public opinion finds Georgia to be the primary enemy of Russia—even more so than the United States. This section will uncover the consequences of the cyber operations conducted during the conflict in 2008. Table 4.2 shows that four cyber incidents have occurred between Russia and Georgia, all immediately before or during the military clashes between the states. The incidents were minimal by cyber standards, as only DDoS and website defacements were employed. As shown in this chapter, Russia had the ability to inflict more damage in Georgian cyberspace; however, it chose not to. Georgia retaliated by flooding certain government sites with DDoS tactics, but the damage was minimal and temporary at best. Russia refrained from the full use of cyber tactics, even in times of physical conflict against its stated enemy.

The first cyber incident Russia initiated on Georgia was a DDoS method on Georgian government websites. The flood of botnets that effectively shut down these sites for days left much of the Georgian population in the dark over what was going on in the separatist enclave of South Ossetia and the immediate border with this pro-Russian region.[57] The Georgian and Russian militaries were engaged, yet many Georgians were unable to get the answers they sought from their leaders because of widespread government network shutdowns. Furthermore, the media were unable to get information, as these reporters also were partially dependent on the Internet.

The VoIP telecommunications network, the largest mobile provider in Georgia, was also shut down in concordance with the conflict.[58] People witnessing the conflict who wanted to get the word out, the media, and private citizens alike could not tell the world what was going on around South Ossetia. The military operation in the remote region was, therefore, not covered with accurate information. Russia was able to swiftly overrun the Georgian military without the eyes of the world watching, and by the time Georgian networks were becoming operable again, Russian troops were 30 miles from the Georgian capital of Tbilisi.[59] The Georgian troops had been soundly defeated, and Russia, perhaps feeling the effects of Western condemnation, stopped short of invading the capital and retreated to its defensive posts in South Ossetia.

The third and final cyber tactic by Russia during the 2008 conflict was a series of vandalizing incidents that supplemented the DDoS incidents on Georgian government websites. Georgian President Mikheil Saakishvalli's website was defaced with portraits of the leader altered to make him look like Adolf Hitler.[60] Similar incidents of this type hit other various government websites. Although rather benign, this cyber weapon had a widespread psychological effect on the Georgian population.

Georgia responded to these tactics with one cyber operation of its own. When word spread that Georgian troops had been killed, the Georgian government, with help from Estonian experts, were able to infiltrate and temporarily shut down several networks used by patriotic "hacktivists" in Russia.[61] These normally offensive hackers were victims of a cyber assault of their own for a few hours, but the tactic was quickly blocked, and Russian hackers were online again to continue assaulting Georgian networks. The cyber incidents during this short conflict left much of the world as well as much of Georgia in the dark about what was happening between Russian and Georgian troops. Russia was thus able to crush the Georgian military and secure South Ossetia without the world watching, and this gave it an advantage in achieving these objectives with ease.

Although more damage was done by Russian guns and bombs during the conflict, Russia did something unprecedented in international conflict: it used cyber tactics in coordination with a conventional military campaign. In its campaigns against Iraq and Libya, the United States has publicly indicated that although it has the capabilities to use cyber tactics that could inflict real damage, it has refrained from doing so in order to not set a precedent where malicious cyber weapons are utilized during a military campaign. Russia, therefore, broke this restraint and could have started a new precedent. However, the cyber tactics used by Russia had no real strategic military value. Their utility was to confuse the Georgian government and have a psychological effect on the Georgian populace.

They were not used to shut down Georgian radar, confuse Georgian troops, or take control of Georgian missile defenses. Furthermore, Russia used these cyber tactics against a Near Abroad state—a region Moscow considers within its sphere of influence. Our theory predicts that Russia would not use this coercive tactic against states with which it does not have an active rivalry and a history of regional control.

When open hostilities existed between Russia and a state considered a rival, Russia refrained from fully utilizing its cyber weapons. The cyber incidents employed during the 2008 conflict were not used as a military strategy and were merely intended to rouse the citizens and leaders of Georgia. Cyber conflict as a Russian tool, therefore, is not essential as a Russian foreign policy instrument, nor is it desired even when Russian boots are on the ground. We assert that Russia is deterred from escalating conflict with cyber tactics, partially because of the backlash it received in 2007 with Estonia and partially because it does not want to be the first cyber power to fully implement its potential in cyberspace. Much like the dynamics of not wanting to be the first state to launch a nuclear strike, Russia is taking a page out of the book of its Soviet past. Russia has engaged in cyber interactions with one more state, its longtime adversary, the United States.

United States

Table 4.2 shows that Russia has infiltrated US networks four times in the period of 2001 to 2013.[62] The first incident involved hacking the personal network of US officials in Georgia in order to cause confusion during the 2008 conflict. This operation's intentions were to delay the American reaction and response to the conflict. The main purpose of this operation was to steal the cyber identities of the American diplomats to cause confusion among Georgian government officials.[63] Although menacing, this incident also only receives a minimal severity score of one.

The next cyber operation Russia imposed on the United States receives a severity score of two. For two days in November of 2008, the US Central Command network for Iraq and Afghanistan in Kyrgyzstan was hacked and secret military information was stolen.[64] Nothing too sensitive was stolen, and a minor row existed between the US and Russia; however, this is nothing new as far as behavior between the longtime rivals. It is also important to note where the incident happened, in Kyrgyzstan in post-Soviet space. The same result from this cyber incident could have been achieved with traditional spying techniques, as it was thousands of times during the Cold War.

The last two cyber operations between the adversaries are when botnets infiltrated the US Eastern seaboard's power grid in 2009 and again in 2013. Although potentially catastrophic, no power was knocked out, and the threat was blocked from doing anything malicious.[65] The source was traced to Russia, but it seems that no ill will was intended, besides perhaps letting Washington know what Russia is capable of in cyberspace. These incidents get a severity score of three each, because of the damage they could have done. It cannot be pinpointed exactly where in Russia the 2009 incident came from, as it seems that Russians are the masters of covering their tracks in cyberspace. The 2013 infiltration was perpetrated by what is now being called the "Dragonfly" group, and at this point little is known about exactly from where in Russia these series of hacks came.[66] Most speculators believe it to be the work of the Russian cyber patriots scattered about Russia.[67]

What can be taken from the observed cyber interactions between the United States and Russia? We assert that these operations must be put into context. First, there are only four cyber exchanges between the rivals over a long time span of ten years. If cyber conflict is to be used to burn the other side, as rivals ritualistically do, we would expect to see more frequency of cyber interactions between the states. Second, both states have the offensive capabilities to inflict more damage on one another. Both sides have refrained from doing so, and we attribute this to the dynamics of restraint due to fear of retaliation. Much like the environment of the Cold War, although both countries have "first strike capability" in that they have the means to inflict real physical damage from cyberspace, neither state wants to be the first to do so out of fear of retaliation.

Third, the cyber interactions between the United States and Russia can be considered part of the normal relations range between rivals discussed in Chapter 3. More recently, cyberspace has been a new arena for an age-old tactic between enemies: spying. Stealing information from or causing confusion and/or panic in your enemy's government and population is nothing new; only the means of achieving this goal is changing as more states continue to increase their reliance upon the Internet for important sects of national security and infrastructure. As the US and Russia, if they wished, could do much more damage in cyberspace than they have to this point, cyber conflict for Russia is not a necessary tactic for achieving national security goals; it is merely opportunistic or supplemental. Based on Russia's capabilities in cyberspace and compared to other states that engage in cyber conflict, Moscow has shown much restraint when using the weapon as a foreign policy tool.

Assessment and Conclusion

Table 4.3 shows how Russia compares with other states that use cyber tactics as a new tool of their foreign policy. As the most cyber-capable state in the world according to our "heavyweight" scale in Table 4.2, Russia is at the bottom of the list in cyber engagement among the most capable states in cyberspace. Therefore, Russia is showing great restraint when using its offensive capabilities as well as successfully defending itself from cyber operations from beyond its borders. China is the most active as well as the most offensive state in the cyber world, with the United States as its favorite victim as well as many East Asian rivals. The United States is ranked second in cyber interactions among the heavyweights; however, American networks are usually the victims, and the US tends to not be the aggressor on the digital battlefield. The interactions among the Koreas are usually with each other, as are the interactions between Israel and Iran. Russia's interactions are with its two rivals, as well as the lone 2007 Estonian cyber dispute.

China is a rising power whose military capabilities dwarf those of the United States. Its economy is also heavily dependent upon and highly intertwined with the American economy. It would be unwise of China to confront the United States militarily, as it would be equally unwise to cut off trading with the United States. How does a rising power, therefore, challenge the reigning superpower for a prominent role globally? We argue that because of these limitations militarily and economically, China has found its niche and advantage over the United States in cyberspace. China is responsible for over half of the recorded cyber operations on the United States, many of which have been able to steal top-secret military strategies and technologies. Furthermore, China is not afraid to

Table 4.3 Number of Russian Cyber Interactions Compared to Other Cyber Powers

Cyber Power	Total Cyber Interactions
China	35
United States	32
South Korea	18
North Korea	15
Iran	13
Israel	10
Russia	9

Source of data: Valeriano, Brandon and Ryan C. Maness. 2014. "The Dynamics of Cyber Conflict between Rival Antagonists, 2001–2011." *Journal of Peace Research* 51 (3): 347–360.

use cyber tactics against its East Asian rivals, who are under the protection of the American defensive umbrella. For China, cyber conflict is its power politics behavior of choice.

As the United States is one of the most "plugged in," technologically advanced, and powerful states in the world, it is no wonder that America suffers the most cyber incidents in the world, which places it near the top of the list of cyber interactions among the heavyweights. China has been able to steal many government and strategic secrets from the free and open American Internet. The United States has yet to find the right balance between security and freedom on the Internet, which for the time being makes the sole superpower particularly vulnerable in cyberspace. It is also very capable offensively, as it is pinpointed as the culprit behind the sophisticated Stuxnet and Flame malware. However, the US has been restrained from escalating its sophisticated capabilities, as it refrained from using cyber techniques in its conventional interventions in Iraq, Afghanistan, and Libya. The United States, therefore, is offensively capable and defensively vulnerable, which has restrained it from escalating behavior in cyberspace with more malicious operations.

The paired interactions of South Korea and North Korea and Israel and Iran are regional rivals in East Asia and the Middle East, respectively. Previous research has found that much of the cyber conflict conducted from 2001–11 has been among states vying for regional prominence.[68] Cyber tactics are supplemental tools in the ongoing quest to burn the other side, which is a key feature of many rivalries. The nearly 60-year ideological rivalry among the Koreas is continuing in cyberspace. Iran has declared a "cyber jihad" on Israel and is helping fringe organizations such as Hamas and Hezbollah conduct cyber malice in Israeli networks. Israel has retaliated in kind to these tactics and is also blamed for injecting the American-made Stuxnet and Flame into the Iranian nuclear network. It is no surprise that these longstanding enemies are using cyber conflict as part of their arsenal.

Returning to our object of focus, what explains Russia's relative restraint in cyberspace? As we assert that Russia's national interests are not global and are mostly limited to its former empire, this same assertion applies in cyberspace. Russia has only engaged Estonia and Georgia in post-Soviet space. The first attempt on Estonia in 2007 resulted in Moscow receiving widespread condemnation from much of the world. Estonia retaliated in the world court of public opinion rather than in cyberspace. It seems this method has worked, as no Russian cyber operation on Estonia has manifested since. Furthermore, it is safe to assert that Russia will probably not engage another Near Abroad state with close ties to the West, as condemnation was widespread over the incidents with Estonia.

The cyber operations on Georgia supplemented Russia's military campaign in 2008, a tactic that is unprecedented in modern warfare. However, we can attribute these incidents to the regional rivalry dynamics discussed above, in that Georgia and Russia are considered rivals and that cyber conflict is not an uncommon tactic between dyads of this type. Furthermore, Russia's capabilities in cyberspace indicate that it could have done a lot worse to Georgian networks, yet it achieved its goals through conventional military means and restrained itself from setting another precedent of escalating the severity of cyber conflict. Similar to the United States' position, Russia does not want to be the first state to fully utilize its offensive cyber capabilities.

We find that the low frequency and severity of the cyber interactions between the United States and Russia is merely a new tool in an age-old technique between adversaries: espionage. The United States and Russia have been rivals for over 60 years, and spying is part of their normal relations range. In other words, these types of cyber incidents are expected between these longtime rivals. Russia, therefore, although quite capable of causing great harm in cyberspace, does not have the incentives other states may have for using cyber conflict as a foreign policy tool more frequently, nor does it receive many outside infiltrations due to tight and sophisticated defense mechanisms.

With the Russian use of coercive tactics against Ukraine in the diplomatic, military, and energy realms as a result of the 2014 crisis, one may expect that Russia may have broken its restraint practices in cyberspace and unleashed its impressive arsenal on its smaller neighbor. In fact, this is not the case. After the Crimean referendum in March 2014, where shortly thereafter Russia officially annexed the peninsula from Ukraine, one media outlet reported that "massive cyberattacks slam official sites in Russia, Ukraine."[69] In fact, this is not the case. These "cyberattacks" were a series of DDoS incidents volleyed between proxies from the two countries that shut down websites for short periods of time.[70] This is nothing new, troubling, or against the international normative practices developing in cyberspace that have been observed so far.[71] Furthermore, at the time of this writing, it is too early to tell whether or not these incidents can be attributed to either the Russian or Ukrainian governments; therefore, official inclusion in this chapter's analysis is thus far unwarranted.

Other media outlets have acknowledged Russia's restraint in cyberspace as the 2014 Ukrainian crisis has continued, and Russia has been implicated in its continuance.

Russia is the most cyber-capable state in the world. It also uses cyber tactics the least among the major cyber powers. Russia's national interests mainly lie in its former empire, with few interests lying outside the region, except to oppose Western influence. To get its way in post-Soviet space, we find that more conventional foreign policy tactics are used in its former Soviet Empire. Outside the region, Russia has been recovering much of its military prowess under the Putin regime. Chapter 3 declares that Russia has no friends except its army and navy, and we cannot expand this phrase to include cyber conflict at the end of this chapter. Russia is restrained in cyberspace, as it is in another area of growing international concern: the Arctic, which is the topic of Chapter 7. The next two chapters cover the most frequent use of power politics in Russia's former empire: coercive energy policy. It is in this issue area where Russia uses its power to a large extent, especially with Ukraine, and it is now known as an energy superpower in the 21st century.

5
Russian Coercive Energy Diplomacy in the Former Soviet Union

Introduction

Among other new forms of power, we have energy power. While not new, the dynamics of the arena have changed. Here we focus on gas and other forms of energy power beyond the scope of oil or nuclear energy. There are new forms of power developing within the energy community, and here we focus on how Russia marshals its newfound power potential.

The Russian annexation of Crimea in March 2014, threatening the sovereignty of Ukraine and alarming policymakers in the West, has brought Russian power politics behavior back into the forefront of debate among media pundits and policymakers alike.[1] Seen as a violation of international law by many Western states, Russia's actions should come as no surprise, if looked at through the lens of the theory of situational coercive diplomacy backed by quantitative analysis.[2] Ukraine and the peninsula of Crimea serve Russia's energy interests, as this part of the former Soviet empire is the most salient to the new Russian state under Vladimir Putin.

Another example of Russian energy coercion with a post-Soviet state is with the small country of Moldova. It is the poorest country in Europe, located at the crossroads of Russia, Ukraine, and the European Union.[3] Dependent on agricultural exports for economic vitality, Moldova has had to make tough choices to provide goods, services, and commodities to its populace. Moldova has frequently made concessions to Russia in hopes of greater access to its markets, and in return Russia would provide essential investments in infrastructure and commodities, specifically in energy. Moldova has also allowed for the relative autonomy of the pro-Russian enclave of Transdniester, a self-governing territory protected by Moscow. Owing nearly $4 billion in natural gas fees to Gazprom, Moldova's intention was clearly to gain favor with Russia; instead the opposite has happened.

Moldova has bent towards Russia in hopes that coercion will decrease; it has sold its majority share of natural gas pipeline infrastructure to Russian energy giant Gazprom in return for below-market gas prices. This "assets-for-debts" tactic has been used not only on Moldova but on many other former Soviet states owing energy debts to Gazprom in order to gain more leverage and have a bigger footprint in these countries.[4] However, Russia has not held up its end of the bargain and has demanded higher prices for energy and has threatened embargoes on Moldovan goods until these concessions were made. In 2006, Russia cut off imports of Moldovan wine, halving the total imports.[5] Russia also demanded an increased price for gas at $400 per 1000 cubic meters, which is even higher than the market price that EU customers pay. It cut off gas supplies in 2006 until an agreement was reached.[6] Moldova has learned, very painfully, that concessions to Russia only result in further demands, something the struggling country cannot afford.

Instead of bending more in the face of continued Russian demands, Moldova and Ukraine have done the opposite and have begun looking to the West. Moldova turned to the EU for help in hopes of detaching itself from Moscow's grasp. In June of 2012, Moldova received a $34 million grant in economic aid from Brussels, and it negotiated with the organization for a free-trade agreement that was signed in July 2014.[7] Ukraine is in the midst of a tug-of-war between Russian and Western influence where blood has been spilled on its soil. It is very likely that the 2014 crisis may push Ukraine towards Western influence permanently, achieving the direct opposite of the strategic ambitions of Putin. Clearly, this is an outcome that was not intended.

As we have outlined in our framework of situational coercive diplomacy, Russia will use its newfound gas power in a coercive manner in post-Soviet space when the issue is salient, the public supports the moves, there is a standing rivalry, and the region in question is important to Russia's great power identity. This chapter will follow this theory to demonstrate that Russia uses its energy power coercively in these contexts. Chapter 6 will show that the Russian public supports Russia's coercive energy policy. This chapter concentrates on a quantitative analysis to uncover which factors correlate with these policies. We will also note how Russia's use of coercive tactics has often led it to fail in achieving its goals. The power politics–style coercive energy policy Russia has implemented is backfiring; Moldova is now moving closer to the West and away from the economic grasp of Russia. Ukraine has also suffered from Russian energy coercion and has also gained a more pro-Western outlook on its foreign policy. Here we examine Russian coercive energy strategies and find that these tactics have similar results in other parts of the former USSR, yet Moscow seems to continue to try to gain leverage on the states of the former Soviet Union either through natural gas subsidies, pipeline transit fees, and the above-mentioned assets-for-debts program, with limited success.

Power, its sources, uses, and potential, has long been at the heart of international relations scholarship. Yet most seem to examine power from the perspective of material power.[8] Others have jumped to examining Nye's "soft" power in all forms, mainly cultural or the power of ideas.[9] What has clearly been lacking in scholarship is the scientific study of the use of the power of resources, the leverage of supply and demand, and the power of commodity accumulation. This chapter will examine how Russia uses modern coercive energy power. Energy is the main engine of power for Russia in its Near Abroad. Our specific research question is how have the Putin/Medvedev Administrations used energy as a diplomatic weapon, specifically natural gas, and which variables influence the Russian government's decisions on these matters? Factors such as close ties to the Russian government, ties to the West, the presence of a foreign policy crisis between pairs of states, whether or not the country is a pipeline transport territory for Russian energy exports, ethno-religious factors, and the amount of ethnic Russians living in former vassal states are all relevant to the examination of Russia's use of energy as a source of power over the customers of its former empire.

This chapter begins with a background of how Russia has become an energy powerhouse in the 21st century. We then move towards an examination of Gazprom, the corporate tool for Russia's natural gas coercive policy. Finally, we present our hypotheses and quantitative analysis that uncovers which factors influence Russia's coercive energy policy. We find that increased Western ties correlate with increased natural gas prices (that are subsidized) for that state the following year. States that have close diplomatic ties with Russia are found to pay the least amount for natural gas. This indicates that Russia uses both the "carrot" and the "stick" with its energy policy.[10]

Russian Energy Power

Power is simply the ability to get someone to do something he or she would otherwise not do.[11] It is the power to coerce and gain leverage. Russia has massive reserves of oil and natural gas, giving it leverage over states in the post-Soviet arena that lack these capabilities. Russia ranks second in oil reserves and first in the supply and reserve capacity of natural gas. This makes Russia an "energy superpower."[12] Russia possesses over 30 percent of the world's working natural gas fields. It is also the second largest oil producer in the world and has nearly 10 percent of the world's reserves, ranking it eighth.[13]

Russia's geographic location makes its resources readily available to important marketplaces willing and ready to pay—Europe, China, and East Asia. Natural gas is only able to travel via pipeline if it is liquefied, and the cost of liquefication is greater than the wholesale and subsidized prices to post-Soviet, Asian, and European markets. Oil, which is always in liquid form, is more easily transported but tougher to refine depending on the location. As Russia has a near monopoly on natural gas in the region, we posit that this form of energy has become the most important form of coercive power in its Near Abroad. Energy power gives Russia the leverage to dominate the region when its rivals challenge Russian desires. Russian domination over its neighbors could in theory expand in scope, pace, and reach as it utilizes this new form of coercive energy power. However, we find that for the most part, this policy is actually drawing post-Soviet states away from Russia and towards the West. Coercive diplomacy, in whatever its form, can lead to negative foreign policy outcomes.

The general turmoil of the 1990s in Russia did not lend itself to the creation of strong institutional development.[14] Instead of creating and stabilizing institutions that would allow for a prosperous democracy, the Russian political elite auctioned off state resources in exchange for cash and political stability. Economic "shock therapy" ended up crippling state capacity and stalling institutional development.[15] Due to the sell-off of assets, power was in the hands of a powerful few, who set up institutions according to their interests. In the 1990s, Russia developed state capacity through institutions that did not have state building in mind but were rather tools of the oligarchs to maintain position and wealth.[16] The economy did not diversify, and energy became the central engine of the Russian economy. Democracy, according to Western standards, failed to take root, and Russia did not meet the requirements for WTO membership.[17] It also had trouble adhering to the Energy Charter Treaty it had signed with the European Union. The treaty states that Western corporations that invest in Russian energy infrastructure and exploration will receive below-market prices for their natural gas. Signed under the Yeltsin presidency, the treaty as seen through the eyes of Putin is a Western attempt to infringe upon Russian sovereignty. The dominant culture of corruption and bureaucracy became the norm, and with no independent judiciary for protection of property rights, foreign direct investment

(FDI) slowed to a trickle.[18] The Russian economy remained stagnant and undiversified, which contributed to the economic collapse in 1998. Therefore, Russia's democratic experiment had already started becoming a façade during Boris Yeltsin's tenure. Putin accelerated this process and consolidated power into the presidency and a few elites in his inner circle. This process did not take much time, as the institutions for authoritarianism were already in place, and Putin just put "the right people in the right places."[19] Russia under Putin has been given many labels to describe its political system, among them are "electoral authoritarianism," "multi-party authoritarianism," "oligarchic democracy," and "petro-state."[20] Property rights and economic clout are given only to those in Putin's inner circle and those most loyal to him. "The assets thus controlled by long-term associates of Putin from his days in the KGB and the St. Petersburg mayor's office include Gazprom, the Rosneft oil company, the Transneft oil pipeline conglomerate, Channel One (the largest television network), railways, a key cell phone company, and the oil-export monopoly. In contrast with the 'oligarchy' of the 1990s, which was open to anyone with enough money, this arrangement is more like a closed janissary caste."[21] Thus the Russian ruling class is under the direction of Putin, and loyalty is rewarded with stakes in Russia's most profitable industries. This structure is quite similar to the Soviet way of thinking, where power is concentrated, and the state controls the key industries.

What Putin has done in the Russian economic sphere is place his inner circle of some former KGB and St. Petersburg contacts in the top positions of the state-owned energy companies.[22] Dmitri Medvedev was a loyal servant during Putin's St. Petersburg mayoral tenure and has enjoyed the CEO position at Gazprom and was named successor to the Russian presidency candidacy for the dominant United Russia Party in 2008. Igor Sechin was also an ally of Putin through their time together in the KGB. Sechin now heads the Russian oil giant Rosneft. Russia's energy recourses and capacity was harnessed and reclaimed by the Russian state under Putin. We now see a concentration of the energy industries in the hands of the Russian government or those willing to work closely with Putin since his rise. The gas giant Gazprom's upper tiers of leadership are filled with "friends of Putin."[23] With the concurrent decline in military power, economic statecraft was able to flourish. Using Baldwin's definition, economic statecraft is the process of "offering economic rewards or withholding economic advantages in order to make other international actor(s) do what they would not otherwise do[; this] means using economics as an instrument of politics."[24]

Russia, under Putin, has become a top supplier of raw materials to the world economy. More than 85 percent of Russian exports come from raw materials or primary commodities. In comparison, the United States' exports total of raw materials and commodities is 26 percent.[25] The oil and gas export is at the core of Russian foreign trade. In 2000, it was 50 percent of Russian export revenue, and in 2008 it climbed to 67 percent.[26] Russia's main source of budgetary revenues is proceeds from oil and gas exports, and 25 percent of tax revenues come from Gazprom alone.[27] Russian economic expansion and the revival of its economy are in large part due to the growth of the oil and gas industry.

This reliance on energy exports leaves the Russian economy very vulnerable. Any movement on oil prices, which global natural gas prices are tied to, has an impact on the current account/monetary policy. A rise in the price of oil means potential outside investors will bet on an improving Russian economy in speculative investment while falling prices will lead to a decline in investments. Oil and gas revenues contribute to macroeconomic stability, but the well-being of the Russian state is always uncertain due to price fluctuations in the external energy market. If energy prices fall to a certain level, this could spell disaster for the Russian economy.[28]

In our framework (Chapter 2), we advanced a theory of situational coercive foreign policy. Russia utilizes its abilities to challenge rivals, those states it has salient issues with, and those states internal public opinion vilifies. Here we examine how Russia has marshaled coercive energy power and commodities income according to the empirical evaluation of outcomes. Russia and Gazprom use natural gas subsidized pricing, pipeline transit fees, and trading debt for infrastructure rights as a political tool to bring allies and enemies closer to its sphere of influence, but what factors determine its use of energy power? This chapter focuses on natural gas pricing only.[29] Gazprom is an arm of the state, and the state's goal is to maximize its leverage and influence in what it perceives to be its exclusive sphere of influence. Russia requires that its former vassal states continue to freely transport natural gas to its customers, the countries of Europe. Just how critical are ties to the West or Russia in determining natural gas pricing models? What influence does conflict have on pricing outcomes? These questions are critical but have failed to be systematically investigated. With this in mind, the following hypotheses will be investigated:

> *H1: Countries of the former Soviet Union whose governments have closer diplomatic ties to Russia than the West will have lower natural gas prices relative to those post-Soviet countries that do not have close ties to Russia.*

H2: *Countries of the former Soviet Union that have experienced a militarized conflict with Russia will see an increase in natural gas prices in the following years.*

Russia's Energy Arm—Gazprom

Gazprom can be traced back to 1965, when Soviet leadership decided to concentrate more on natural gas production and consumption.[30] Called the Soviet Gas Ministry or the Ministry of the Gas Industry at the time, the government entity skyrocketed in importance with the discovery of huge reserves of natural gas in the Volga River region, the Ural Mountains, and Siberia during the 1970s and 1980s.[31] In 1989, with the reforms of "glasnost," or openness and transparency under Gorbachev, the Gas Ministry became the first state-corporate enterprise in the Soviet Union and changed its name to Gazprom.[32] Victor Chernomydrin became the first head of the new entity. Although still controlled by the state, stocks were sold so that more people in government had control. When the Soviet Union fell in 1991, Gazprom lost its control in former Soviet space and only retained control of assets in Russia.[33]

In 1993, President Boris Yeltsin appointed Chernomydrin prime minister, and Gazprom began to gain political influence.[34] Yeltsin's government began what has been famously known as "shock therapy," or the rapid privatization of Russian industry and infrastructure. By 1994, 33 percent of Gazprom's shares were owned by the Russian public, 15 percent of the stock went to employees, with 40 percent of the stock retained by the state.[35] In 1996, a small percentage of Gazprom's stocks were sold internationally. By the year 2000, however, as Chernomydrin was prime minister and held a vested financial interest in Gazprom, the company evaded taxes and conducted widespread asset stripping with the assets going to board members and their families. The Russian state received next to nothing in tax revenues and dividends, and the Russian oligarchs of the 1990s got away with stripping Russia of all of its resource assets.[36]

When Putin became president in 2000, he returned control of Gazprom to the Russian state. One of Putin's goals was to rein in the oligarchs, whom he saw as mismanaging and looting Russia's industry. He fired Chernomydrin and replaced him with St. Petersburg political ally Dmitri Medvedev, the future president of Russia. He helped stop the asset stripping and also regained much of what was lost. Arrests were widespread, and by 2005 Gazprom was in the hands of the state and those loyal to the Putin Administration.[37] Fifty-one percent of the company is now owned by the Russian state, and in 2006 Gazprom was given the exclusive right to export natural gas outside of Russia.[38] Since 1991, Gazprom has been the largest firm in Russia.[39] Gazprom almost constitutes a Russian natural gas monopoly and is also its fifth largest oil producer. It produces over 83 percent of Russian gas, controls the domestic pipeline system and Russian gas exports, and has ownership in pipeline infrastructure in many former Soviet countries.[40] Gazprom also controls many of the banks, industries, farms, and media conglomerates in Russia.[41] Ownership of these various industries gives Gazprom and subsequently the Russian state great reach in keeping the public supportive of state policies.

Gazprom is highly integrated with the federal state elite. It is a remnant of Soviet times, and its operations were left intact after the end of the Cold War. Its board members and financial officers are handpicked by Putin's inner circle, and the Russian government owns over 50 percent of company stocks. Non-Russians are usually not allowed to own shares, ensuring that its function of advancing the national political agenda of Russia's elite remains intact. The largest foreign investor is German-based EON, with a 6 percent stake.[42]

Pipeline investments and ownership as well as foreign debt in the former Soviet Union have large geopolitical implications. As Gazprom owns production and transport mechanisms, as well as being under the control of the Russian foreign policy elite, Russia has the unique ability to use natural resources as a foreign policy tool. This gives it substantial leverage over consumers, especially those who are 100 percent dependent on Russian gas.[43] Rosneft is the oil counterpart to Gazprom, and Transneft is the oil pipeline counterpart. Rosneft is the oil company that, with Putin's crackdown on the oligarchs, gobbled up the assets of the other Russian oil companies of the 1990s.[44] It is headed by Igor Sechin, one of Putin's loyal followers dating back to their time in the KGB. Since these state takeovers, many have argued that Moscow has been using its energy endowments as political weapons, especially against the states of its former Soviet empire.[45]

Ukraine, the largest of the post-Soviet countries and also the most important gas-transit state for Moscow, has been less than cooperative over the years, and this culminated into a domestic crisis in 2014, where the fate and sovereignty of Ukraine is still very much in question at the time of writing. Most Russian natural gas headed for Europe travels through Ukrainian territory, and over 66 percent of Ukraine's domestic natural gas consumption comes from Russia.[46] Ukraine has, however, a foreign policy tool of its own to counter this dependence, as the main oil and natural gas pipelines to Central and Western Europe pass through the former Soviet republic. Its gas pipeline system has a capacity of 120 billion cubic meters per year, and Ukraine receives transit fees from Gazprom, paid in gas subsidies and cash.[47] According to a report in 2006 by the International Energy Agency, 84 percent of Russia's gas exports and 14 percent of Russia's oil exports transit through Ukraine's pipelines, most of which are going to Russia's most important and high-paying European customers.[48] Therefore, Russia's decision to cut off gas supplies to Ukraine is clearly political rather than economic, and it potentially hurt Gazprom's reputation as a reliable gas supplier to its European customers.[49] Russia believed that Ukraine was siphoning off gas from its pipeline without permission and owed Gazprom money. The mid-winter gas shutdown was to coerce Ukraine into cutting a deal to repay its debt. Another factor was that Ukraine had elected a president who had anti-Russian feelings—Yushchenko in 2004—and this was a political move in reaction to the new president's election.

Russia implements a policy of "dual pricing" by energy giant Gazprom for natural gas, which means that Gazprom charges foreign customers more than domestic ones. It fulfills much of the European continent's natural gas needs—effectively 30 percent of the EU market.[50] A disproportionate amount, two-thirds, of Gazprom's revenues comes from its European customers. This allows for the subsidization of domestic prices at home. Although not profit maximizing, it helps the domestic constituency remain loyal to Putin since prices are so low.

With Gazprom's peculiar pricing system, Russia uses the energy dependence of many post-USSR states to limit sovereignty and desires to orientate towards the West. The goal is to have political dominance of the region. Ukraine has seen its gas supplies cut off twice—once in 2006 and again in 2009—both in the midst of winter. Estonia saw its Russian oil supplies suspended in the Bronze Statue dispute with Russia, which also resulted in widespread cyber attacks—the topic of the previous chapter. Moldova has given in to demands from Gazprom and the Russian state, only to be met with more demands and less control over its energy infrastructure. What factors decide how Russia rewards or punishes its gas-dependent former vassal states?

Energy and Coercive Diplomacy

Europe pays the market price for Russian natural gas, and the Russian public pays a very low subsidized rate. Gas prices for post-Soviet spaces are also subsidized and usually lie between the European market price and the Russian domestic price.[51] Energy, for the purposes of this chapter, is the commodity of natural gas. Although Russia uses its oil reserves, pipeline ownership, and debt leverage for political statecraft in post-Soviet space, it is natural gas that Russia has the most control over and a near monopoly of supply to many of its former satellites. Natural gas is Russia's primary tool for economic statecraft. Russia exerts its political muscle by the means available when interests clash on salient issues. Russia dwarfs these states in land area, population, and resource endowment, thus it is the most powerful state in the post-Soviet region, and it has the ability to use coercive diplomacy when there are issue disagreements. As Russia tries to meet its great power identity in the post-Soviet period, it is only natural that it exerts the power of natural gas endowment on unfriendly states such as Georgia and Estonia.[52] This is exactly what has happened in Ukraine, with gas price disputes between Russia and Ukraine leading to two pipeline shutdowns during winter in 2006 and 2009.

We argue that certain contextual factors decide how much Russia charges its natural gas-dependent consumers in order to bully its former Soviet empire to adhere to the interests of Moscow. The issue of energy is very important to Russia in terms of its stakes as well as its image of itself as a great power. When the stakes are both of a concrete and a transcendent nature, a very peculiar pricing system emerges in the Russian Near Abroad.[53] To continue its domestic subsidized pricing system, Gazprom needs foreign revenues from the EU, Near East, and Asia. To get its product to these important customers, it needs the states of the former empire to be reliable and complicit transporters. The prices it charges these states for their natural gas is dependent on the level of rivalry, the issues under disagreement, identity issues of the three sub-complexes, and public opinion. Chapters 3 and 4 show where Russia holds its most intense rivalries, and Chapter 6 will cover the salience of the energy issue and the views of the Russian public; therefore, this chapter focuses on the issues at stake.

Russia is attempting to rebuild a sphere of influence in Eurasia to counter the United States and the West's growing activities in what Russia considers its exclusive political and historical territory. We argue that ties to the West, a key symbolic issue of disagreement, will have an influence on the economic statecraft that Russia imposes on its Near Abroad neighbors. This chapter will only consider natural gas prices, as many states once part of the USSR are completely dependent on Russia for their natural gas needs, thus giving Moscow a greater opportunity to coerce them into political adherence.

Russian Natural Gas Statecraft

Examples of natural gas economic statecraft by Russia towards its former empire are plentiful. With regards to the Baltic area, evidence of Russia's maximization of influence on these states lies with the Nordstream pipeline project.[54] Nordstream is a pipeline that is being built under the Baltic Sea directly to Germany, Russia's best European customer. The pipeline is funded by Gazprom and several German multinationals. Russia has faced opposition by Ukraine and Belarus in attaining full control of energy infrastructure (mainly the pipeline system) in Central and Eastern Europe. Russia is trying to entirely bypass countries in the region in order to avoid the political complications that come when trying to use its energy leverage towards these states.[55] This pipeline would have a capacity of 55 billion cubic meters per year, and it is now fully operational.[56] However, many states in the region have objected. Cheaper land-based pipelines could be built through Estonia, Lithuania, Latvia, and Poland, or gas could continue to be transported through existing pipelines in Belarus or Ukraine. Why opt for the expensive and logistically difficult underwater pipeline? This plan would give Russia the maximum strategic advantage to subordinate its former satellites. Some central former Soviet states such as Ukraine and Belarus are transit countries, where pipelines of Russian gas flow through their borders on its way to Europe. These transit countries are politically troublesome since they can impose demands on Russia rather than Russia imposing demands on them.

In 2007, Gazprom and the Italian firm ENI signed a joint agreement to build what is now popularly known as the "South Stream" gas pipeline intended to run through Russia to Turkey.[57] The pipeline would then traverse through the Balkans, with branches to Austria and Italy. Bulgaria, Serbia, and Hungary are also signatories to the project. In early 2015, however, the South Stream project was halted due to low oil and gas prices as well as the U.S.-led economic sanctions imposed on Russia for its invasion of Crimea and subsequent involvement in Ukraine's domestic affairs.[58] South Stream would also bypass the current transport states of Belarus and Ukraine. This pipeline, like the Nordstream project, is intended for the EU market and will not serve any new customers. As the same gas could reach these markets through Ukraine and Belarus, why spend billions on a pipeline that serves no new customers? Some argue that it is about market share and the reduction of the possibility of future competition with potential Western investors.[59] This will allow Russia to maintain its current market share and also deter others from attempting to penetrate the lucrative EU market. Furthermore, the pipeline will also allow for Gazprom to take transport privileges away from Ukraine, the state which has caused the Russian elites much grief over the years with the 2006 and 2009 gas disputes. With more options to circumvent Ukraine and Belarus, Moscow will finally be able to keep Kiev's and Minsk's leverage over Gazprom in check. With the annexation of Crimea by Russia, this area could serve as a more affordable route for the South Stream project, as less pipeline would have to be laid underwater if the Crimean landmass is utilized. It remains to be seen as to how Crimea will be used by Gazprom and Russia for energy purposes. This gives Russia greater ability to utilize economic statecraft without being checked by external powers.

Table 5.1 gives the 2005 prices for most former Soviet countries.[60] In 2005, Gazprom came under the control of the Russian state under Putin; the table shows the immediacy of his administration's coercive energy policy. There is no market logic behind the differing prices, nor do the subsidies make any superficial sense. Gazprom's official reasoning for the differing prices in post-Soviet space reads as, "In its relations with the CIS and Baltic states, Gazprom adheres to the policy of phased transition to the contractual terms and conditions, and the pricing mechanisms for gas delivery and transit services similar to those applicable in the European countries."[61] This statement really tells us nothing of the nature behind the pricing. From Table 5.1 we can gather that Estonia pays the highest price for its natural gas from Russia ($95 per cubic meter in 2005). This is not surprising since Estonia is also the most hated state

Table 5.1 Comparative natural gas prices of the former soviet union in 2005

Near Abroad Country (Region)	Import of Natural Gas ($mil)	Share of Russian Gas (%)	Average Price of Russian Gas ($/cm)
Armenia (Caucasus)	91	100	67
Azerbaijan(Caucasus)	277	9	45
Belarus (Central)	950	100	48
Estonia (Baltic)	91	100	95
Georgia (Caucasus)	86	100	73
Latvia (Baltic)	151	100	95
Lithuania (Baltic)	264	100	90
Kazakhstan (Central Asia)	372	46	33
Moldova (Central)	108	44	80
Ukraine (Central)	3946	31	52

Source of data: Ivanenko, Vlad. 2006. "Russian Energy Strategy in Natural Gas Sector." Accessed 4/13/2011, available at: http://papers.ssrn.com/sol3/papers.cfm?abstract_id=953467.

by Russians according to public opinion polls (see Chapter 3, Table 3.2). Estonia has also achieved outside alliance membership in NATO and the EU.[62] Estonia is now under the military protection umbrella of the power of the West. If Russia, the country that sees itself as the dominant power of its region, can no longer challenge Estonia militarily, it will do so economically through its energy might.

Attempts at energy statecraft have been particularly felt in the European post-Soviet states of Ukraine and Moldova. These states have suffered natural gas shutdowns through their pipeline systems when disagreements over pricing, pipeline access rights, and possibly attempts at democratization or a closer alignment with the West occur. Ukraine is a special case as it is the busiest natural gas pipeline transit country for Russia. Russia and Ukraine are constantly renegotiating gas and transit prices, respectively. The fact that Ukraine has attempted alignment with the West and has attempted to integrate itself into the European Community several times may have led Russia to rethink these gas and transit pricing agreements (although in 2009 a more pro-Russian government was ushered in to power). The 2014 crisis is partially a result of the latest fallout between Kiev and Moscow over issues of Western or Russian allegiance. Thus Ukraine has been a centerpiece for contention among Russia and the West, and perhaps this pipeline leverage is another reason why. Ukraine has taken a stand against Russian attempts to dramatically increase gas prices, even in the light of shutoffs in the midst of winter (2006 and again in 2009). For now, Ukraine has been troubling for Russia, as it has great leverage as the major transport point of Russian natural gas.

On 2 October 2007, as it was becoming apparent that Orange parties were going to control the Ukrainian Parliament, Gazprom again began economic statecraft with Ukraine. It threatened to reduce gas supplies by 50 percent if Ukraine did not pay its outstanding debts to Gazprom by the end of the month.[63] This crisis was averted when Kiev agreed to a natural gas price of $179.50 tcm for 2008, a 38 percent pricehike compared to 2007.[64] However, this increase was double that given to Belarus, another transport country with anti-Western, pro-Moscow sentiments.

Another Russo-Ukrainian gas crisis began on 1 January 2009, when Gazprom stopped gas supplies to Ukraine. The issues that led to this gas row included debts owed by Ukraine to Gazprom and disputes over the price that Ukraine would pay for 2009.[65] The shutoff was intended only for the gas for the Ukrainian market, as Russia continued to send gas intended for its European customers. When these supplies did not arrive, Russia accused Ukraine of siphoning off gas for its own use, and by January 6 it cut off all deliveries through Ukraine, which undersupplied the rest of Europe in the midst of a very cold winter.[66] On 18 January 2009, Gazprom and Ukraine came to an agreement, and gas transport to Europe resumed on January 20.[67] All of these rows and disagreements over gas occurred under the Yushchenko presidency, which was boisterously pro-Western and anti-Russian.

The 2009 contested elections (because of possible election rigging supported by the Russians) of pro-Russian Yanukovych should lead to an even cheaper natural gas price, according to our model. The Nordstream and South Stream pipelines are clearly Russia's future tool to regain leverage with gas in the region. However, if the Putin Administration can rein in the government of Ukraine to remain friendly to Moscow and more hostile to the West, there may not be a need for continued expensive construction for these logistically difficult pipeline networks.

The 2014 crisis in Ukraine that has been tearing that country apart also has its roots in economic issues where Russia and the West have been divisive. In November 2013, in an unexpected move, then-President Yanukovych rejected a free-trade economic deal with the EU.[68] Thousands of Ukrainians took to the streets in protests, seeing the move as a result of strong-arm tactics from Russia.[69] Indeed, the following month, in December 2013, the Russian response to the rejected EU deal was a bailout package and economic deal of its own. The deal was worth nearly $20 billion and was to ensure that Ukraine remained in Moscow's political orbit for decades to come. It forgave the $5 billion debt already owed by Ukraine to Russia and also included a steep decline in the price of natural gas for the country.[70] As a result of this deal, the protests in the streets of Kiev and other Ukrainian cities escalated and turned violent, which eventually led to the ouster of Yanukovych and his subsequent exile in February 2014. The new government in Ukraine, under the leadership of Petro Poroshenko, has rejected this Russian deal, and the country is now on track to economically integrate with the EU and West. On 24 June 2014, Ukraine signed an association agreement with the EU along with Georgia and Moldova.[71] However, it seems that Russia will not let Ukraine go without a fight, as ethnic Russian separatists in the Eastern part of Ukraine are destabilizing the country, leaving its future very much in question.

Moldova has also suffered crises with Russia similar to those of Ukraine. On 1 January 2006, parallel to the Russo-Ukrainian dispute, Gazprom cut off natural gas supplies to Moldova when Moldova rejected Moscow's doubling of the price it charges Moldova for natural gas.[72] Gas supplies were restored on 17 January, when Moldova agreed to a natural gas price increase from $60 per thousand cubic meters (tcm) to $110 tcm. Moldova also agreed to give Gazprom its 13 percent stake in MoldovaGaz, the overseer that controls Moldova's pipeline and energy infrastructure, in return for Gazprom forgiving some of the $4 billion debt.[73] As Moldova is also pro-Western and democratizing, Moscow saw the need to punish it for its allegiances. Unfortunately for Moldova, it does not serve as a courier to Europe as Ukraine does, thus it has no ability to counter Russia's coercive energy power.

Even Belarus, considered to be Russia's closest and most supportive ally, was the victim of Russian energy coercion in the winters of 2004 and 2006–07.[74] Belarus, as a reward for its adherence to Russian foreign policy, has enjoyed subsidized near Russian-domestic prices since independence in 1991. Where friction has occurred between the countries is the Russian reluctance to allow Belarusian manufactured goods in exchange for Belarus's cheap natural gas pipeline fees, a pipeline that also feeds the energy needs of the Russian enclave of Kaliningrad, which has been cut off from mainland Russia since the Soviet breakup.[75] Belarusian President Lukashenko, therefore, decided to play this economic card against its more powerful neighbor, once in 2004 and again in 2006–07.

The dispute in 2004 began over a disagreement over the 2002 agreement that Belarus would continue to receive Russian domestic natural gas prices as long as Gazprom was given 50 percent control over Beltransgaz, the Belarusian company in charge of the natural gas pipelines within its territory. Gazprom would therefore be able to more reliably supply Kaliningrad and its Western European customers.[76] President Lukashenko disagreed with the price that Gazprom was to pay for its stake in Beltranzgaz (Gazprom wanted to pay $1 billion, but Belarus was asking $2.5 billion), and for this difference in opinion, Gazprom immediately raised the prices on gas. Neither side budged; thus, on 1 January 2004, Gazprom shut off all deliveries to Belarus, which also meant that gas deliveries did not make it to European customers. Gazprom faced international backlash and quickly reestablished gas supplies through Belarus. Although Gazprom was able to raise prices slightly, it did not achieve the desired 50 percent control over Belarus's pipelines. Having not forgotten this incident, Russia countered with another gas dispute in the winter of 2006–07.

In March 2006, Gazprom announced that Belarus would begin paying closer to market prices beginning in 2007.[77] Belarus then announced that it would raise the transit fees on both natural gas and crude oil if these prices were to increase. Gazprom countered by offering the original $2.5 billion for its 50 percent stake in Beltransgaz, but Belarus would have to pay a higher price for natural gas, albeit well below the European price.[78] In the end, both sides got what they wanted; Gazprom got control of Belarusian pipelines, and Belarus continues to receive subsidized prices for its political loyalty to Moscow. Regardless, here is evidence that Russia is willing to coerce even its closest friends if they are perceived to be asking too much and attempting to counter Russian power and dominance in the European sub-complex of post-Soviet space. It seems that building pipelines that circumvent these post-Soviet countries is the future tactic for energy coercion. This topic will be covered in more detail in the concluding chapter.

There has also been talk of the West's slow reaction to recent moves by Putin and his energy policies.[79] Baldwin posits that states offer economic rewards or withhold economic advantages, using economics as an instrument of politics.[80] This suggests that the European Union has little leverage against Russia, as it is dependent on Russia for its gas needs. The United States, which has required cooperation with Russia in regards to the use of post-Soviet space for transport and bases for its war against terrorism in the Middle East and South Asia, has also been slow to reply. Both the EU and the United States are scorned for their lack of tough diplomacy in regards to Russia's recent actions against Ukraine, Moldova, and Georgia.[81] These arguments contribute to our study as they suggest that Russia will bully its former satellites without adequate counter-diplomacy by the West.

Georgia's pro-Western path began after the Rose Revolution of November 2003, which ousted the pro-Putin government, even after accusations of widespread use of electoral fraud to try to hold on to power were made.[82] In 2005, after nearly two years of democratic reform and closer ties to the West by President Mikheil Saakashvili, Gazprom announced price increases for natural gas headed for Georgia. In November 2006, it went even further and declared that it would cut off gas supplies to Georgia by the end of the year unless Georgia agreed to a 100 percent increase for the price of gas or sold its main pipeline to Gazprom.[83] Economic statecraft for Georgia, due to its pro-Western orientation, is therefore apparent. However, Georgia's leverage against Russia lies with the fact that it shares a border with energy-rich Azerbaijan. This has allowed it to counter Russian economic statecraft, as Georgia does have options for where it gets it natural gas. The closest alternative is Azerbaijan and the Baku-Tbilisi-Erzurum (BTE) gas pipeline that circumvents Russia to Turkey. Yet Georgia has fought a major military conflict with Russia (2008) and has held close ties to the United States. It is Tbilisi that pays a disproportionate price for natural gas when compared to other Caucasus states, yet the BTE pipeline has given it an option.[84]

Coercive Hierarchies

David Lake's theory of hierarchy for powerful states is useful in explaining how Russia interacts with its former empire.[85] Rooted in Kindleberger and Krasner's hegemonic stability theory, Lake assumes that systemic anarchy can be tamed by hierarchies of powerful states.[86] Hegemonic stability theory argues that with the presence of a hegemon, or leader, the collective good of free trade and stability will flourish. The leader coerces its subordinates into compliance with its vision of stability, and these subordinates are basically forced to comply because of power considerations. For Lake, hegemonic stability theory is theoretically sound and an empirically valid theory of modern institutionalism; yet it misses one critical factor that led to its possible empirical anomalies: coercion. Lake suggests the subordinate states must willfully accept the leadership of the hegemon. It is in the self-interest of minor states to follow the leader of the hegemon because the relationship is reciprocally beneficial.

The strength of a hierarchy is dependent upon the adherence to norms and institutions of the hegemon. Russian-backed norms and institutions are not universally accepted, and this leads to a gap in application. Some states, such as Russia, will use coercive diplomacy to achieve foreign policy goals in a regional hierarchy. Norms and institutions fail in regions that have a proliferation of rivalries and highly salient issues, and it is through this lens that our theory of situational coercive diplomacy can be combined with regional hierarchies to explain how states act the way they do in a regional subsystem. Coercive economic statecraft, resentment, and bullying become the norms for Russo–post-Soviet state interactions because of historical patterns or active disagreements.

Research Design: Russia's Energy Policy in the Former Soviet Union

The previous sections have led to the quantitative analysis of Russia's coercive energy policy conducted here. The qualitative examination of the issue is presented in the previous chapter and is useful, but we can only understand the full patterns, interactions, and effects on the process with a quantitative study. This section uncovers the factors that lead to either lower or higher natural gas prices according to an analysis of all cases in order to eliminate some potential biases and confounding explanations. As hypothesized earlier, we assert the factors to be diplomatic closeness to Russia; whether a militarized interstate dispute (MID), which is a militarized display by one state against another that falls short of the empirical definition of war's 1,000 battle death minimum,[87] has occurred between Russia and each state; and present or past ties to the West that are correlated with the price each state pays Gazprom for its natural gas imports.

Impact Factors

Belarus and Kazakhstan each have authoritarian regimes that side with Russia. It is no coincidence that these states pay the least for their Russian natural gas. These authoritarian regimes give Moscow reliable allies and are rewarded with stable and cheap gas. Ukraine also pays relatively very little, and we attribute this to its uniqueness as the courier for the majority of Russian gas to Europe and less so to its pseudo-democratic government.[88] The Baltic States, Moldova, and Georgia all pay higher gas prices for their freedom from Moscow's political grasp. Moscow can rely less on compliance from these democracies, which may elect pro-Russian accommodationists or anti-Russian hardliners. The inconsistency of leadership preferences, we argue, leads to colder relations and higher gas prices with Russia.

Operationalization of the diplomatic ties to Russia comes in two forms. The first are categorical variables replicated and extended from the Diplomatic Correlates of War (DCOW) project.[89] The variables are coded based on the amount of diplomatic relations among dyads for every five years. For the purposes of this study, we extended the coding rules for DCOW and gave countries scores for each year in the 2000–10 range. The dyads consist of Russia and each country in the analysis. Each country in the dyad is assigned a score based on the level of diplomatic relations for each country and year:[90] "0" for no evidence of diplomatic exchange, "1" for *charge d'affaires* representation, "2" for minister-level representation, and "3" for ambassador status.[91] For this analysis, I simplify the coding and code Russia and each country as "1" if there are diplomatic exchanges at any level and as "0" if there is no diplomatic exchange. This has happened often between Russia and post-Soviet states. We expect that countries that hold diplomatic exchange status (on both sides) will pay the least amount for their natural gas, while countries with no diplomatic representation will pay the most for their lack of cooperation with Russia.

The second variable captures whether or not a militarized interstate dispute has occurred within each dyad. These are found by examining the Correlates of War (COW) dataset in MIDs.[92] Here we hypothesize that because of Russia's great power identity crisis, it will not let recent past animosities go very easily. Therefore, if a militarized dispute has occurred between Russia and a post-Soviet state, Russian coercion in the form of higher gas prices will be the result. Examples in this analysis include the Russo-Ukrainian territorial dispute over Tuzla Island in 2003 and the Russo-Georgia conflict in 2008. The operationalization is a dichotomous dummy: "1" when an MID has occurred between Russia and each state, "0" when no dispute has occurred.

The next variable predicts that remnants of Cold War mentalities are still in the upper echelons of Russian leadership. Post-Soviet countries that attempt cozier relations with the US, EU, and other Western countries at the expense of their relationship with Russia will pay more for Russian natural gas than those states that have closer ties with Russia. Georgia has good relations with the US and has paid the most for natural gas in the Caucasus region. Since the fall of the Soviet Union, Ukraine's regimes in power have fluctuated between East and West, and Russia has cut off gas supplies during anti-Russian government control. This leads us to hypothesize that ties with the West may increase natural gas prices. One of the stated goals of the Medvedev and Putin governments is to balance the West. Former Soviet countries that hinder this balance pay the price. This variable is operationalized in two ways: if the state has attained membership in the Western military or economic alliances of GUAM, NATO, and the EU or its Neighborhood Policy, or if it has expressed a serious desire to do so in the past five years. For this requirement, membership ascension to the EU, its Neighborhood Policy, GUAM, and NATO is documented. Western ties are coded as "1"; if the state's government is more Russia-friendly during different time periods of 2000–11, this is coded as "0." This factor fluctuates with regime and foreign policy changes over time. Many countries, such as Ukraine, Georgia, Uzbekistan, and Azerbaijan, have switched their ties between the West and Russia. Before looking at the results of our analysis, a description of several control variables must be discussed.

Regional controls are also added to the dataset to control for the differing energy salience Russia has with each of the three regional sub-complexes of post-Soviet space: European, Caucasus, and Central Asia. Because the European and Caucasian sub-complexes are more salient to Russia and its great power identity, it is important to control for these regional factors. A "1" is indicated for each dyad and region of the post-Soviet state, "0" otherwise.

The pipeline variable looks at whether or not the post-Soviet state serves as a courier for Russia's natural gas to other markets or sells its gas to Russia for export to external markets. Belarus and Ukraine are examples for the former, Kazakhstan and Uzbekistan for the latter. The prices of the courier and non-courier countries seem to differ without any market reasoning. Belarus pays about the same as Ukraine (courier) and Azerbaijan (own gas supply and alternative courier) yet is 100 percent reliant on Russian gas. Belarus, however, serves as a courier to Europe, which may be why the state receives a discount. Therefore, we argue that good relations with Moscow are another factor deciding the price of gas for states, and this variable has already been operationalized above. This variable is coded "1" if the state serves as a courier for Russian natural gas, "0" otherwise.

We also control for the number of Russians living in the country of interest. Russia has granted citizenship to all Russians living in its Near Abroad countries. If something goes awry, Russia has reserved the right to protect its citizens in those countries, including invasion. If ethnic Russians are mistreated in any way, Russia has retaliated (Estonia, Ukraine, Georgia). If these countries, therefore, treat ethnic Russians favorably, Russian gas prices should be low. We code this variable as "1" if at least 15 percent of the state's population consists of ethnic Russians, "0" otherwise. Fifteen percent is the average amount of Russians who live in all countries of the former USSR; therefore, countries that have more than 15 percent are above average, and this serves as a logical cut point. We expect that countries with high Russian populations should pay less for their gas and those with little to no ethnic Russians to pay more.

The dataset for these analyses is comprised of 14 Russia-Near Abroad state dyads from 2000 to 2011. We delineate these dyads into separate groups, and as the analysis is over a period of time, we use the most appropriate technique of panel data regressions. Panel data is used to observe the behavior of different entities across time. Here, the entities are dyads, and we look at effects of different contextual impact factors as well as controls for these pairs of states to get an overall analysis of Russia's coercive natural gas pricing system. For these purposes, we measure the effects of these factors on the different natural gas prices for all dyads. There are two models that can be used to uncover these effects using panel data: random effects and fixed effects. Random effects models assume that the variation across our dyads is random and uncorrelated with the independent variables in the model. Random effects are useful if it is believed that differences across dyads have some influence on the dependent variable. For this analysis, we assume that the differences in the nature of each dyadic relationship will have an influence on natural gas pricing. The intensity and relations range for each dyad are not the same; therefore, the random effects model is appropriate for the analysis. However, we find it appropriate to run the fixed effect model as well.

Fixed effects models are used when the primary interest is analyzing the impact of variables varying over time. This model assumes that each dyad has its own individual characteristics that may influence the independent variables. Something within each dyad may affect either the independent or dependent variables and must be controlled. Another assumption of fixed effects is that each dyad is different and thus the error term and constant of each dyad should not be correlated with the other dyads. As we are also interested in the separate effects of the impact factors and natural gas pricing, we also run a fixed effects model that treats each separate dyad as a dummy variable on the others. Only the main independent variables of "Russian ties," "post-Soviet ties," "Western ties," and "MID" are used in this model, as the dyadic dummies already control for the attributes of the others. We therefore run both models for panel data. Random effects are run to get an overall picture of the contextual variables on natural gas pricing on the entire population, and fixed effects are run to uncover the individual and unique effects of the variables and natural gas pricing for each dyad in the dataset. Next we explain the nature of the dependent variables in the data, as well as the unit of analysis.

Data Analysis: Russia's Energy Bully Pulpit

The unit of analysis for these quantitative analyses is dyadic week. The time period analyzed is 2000 to 2011. The dependent variable is lagged one week for these models. For the dependent variable, collection of different natural gas prices from different regions came from a number of sources, including the International Monetary Fund (IMF), the Oxford Institute for Energy Studies, and the United Nations Commercial Trade Statistics Division (UN Comtrade). These prices come in different weights and currencies; therefore, we convert all prices to US dollars per million cubic meters ($/mcm) for these panel regressions. The gas-scarce Central Asian states of Kyrgyzstan and Tajikistan are not included in the analysis because these states have no direct pipelines from Russia nor do they trade with Russia for natural gas. Therefore, there is no pricing data for these states with Russia or Gazprom. These states receive their gas from neighboring Uzbekistan and Turkmenistan and are beyond this chapter's scope. For the European and Caucasus regions, the prices in the model are what each state pays Gazprom for natural gas. The

Central Asian states of Kazakhstan, Uzbekistan, and Turkmenistan are coded differently but at the same measurement in dollars per million cubic meters. The prices for the dependent variable for these states are what Gazprom pays for their natural gas. The same concept of Russian coercion applies, however, as if these states were to fall out of Moscow's good graces, Gazprom would respond by paying these countries less for their natural gas, as the choices for routes to external markets have been under their monopolistic control until recently, with China's entry into this part of the world.

Tables 5.2 and 5.3 show the results from several panel regression analyses. Table 5.2 shows the results of the random effects model, which controls for differences between dyads and the effects on the dependent variable. The "post-Soviet" relations diplomatic coefficient shows negative statistical significance. When a state has close diplomatic ties to Russia over time, it is rewarded with lower natural gas prices. It seems that changing diplomatic relations and foreign policy crises are correlated with fluctuations in the price Russia charges for gas, at least when the state has kept embassies running with Russia throughout the entire time period studied. Russian diplomatic relations do not have statistically significant effects on the price it charges its post-Soviet customers for gas. This finding shows the importance of contemplation of friendly relations with the West. For the diplomatic variables, Azerbaijan, Georgia, and Ukraine have recalled their ambassadors from Moscow within the time frame of this analysis, and it seems that this variable is a proxy when it comes to the price they pay for Russian natural gas. Belarus,

Table 5.2 Russian economic statecraft with natural gas prices ($ per mcm)

	Coefficient	z-score
Russia ties	0.110	1.30
Post-Soviet ties	−0.453***	−5.61
Western ties	0.094*	2.01
MID	−0.056	−0.92
Europe	Omitted	–
Caucasus	−0.240**	−3.15
Central Asia	−0.743	−0.17
Pipeline	−0.451***	−6.22
Ethnic Russian	0.133*	1.90
Constant	−0.082	−1.50

$p < 0.001$***, $p < 0.01$**, $p < 0.05$*

Source of data: IMF (2013), UN Comtrade (2013), Henderson et al. (2013)
N = 1206

Armenia, and the Baltic States have all kept full diplomatic relations, thus this factor has not affected or lowered the price these states pay for their heat for the winter months. Russian energy coercion, therefore, is at play in the diplomatic realm of its energy policy.

Belarus and Kazakhstan are authoritarian states that have remained loyal and obedient to the Putin and Medvedev Administrations since their rise to power in 2000. Belarus has enjoyed low gas prices throughout the period of this analysis. Belarus also enjoys close trading relations with Russia, as Russia buys up many of Belarus's surplus agricultural products in exchange for cheap gas prices as well as unabated transport of Russian gas to European markets. Similarly, Kazakhstan enjoys a good and cost-effective relationship with Gazprom and Russian leadership. Kazakhstan has its own endowments of natural gas, yet it needs Russian pipelines to transport to foreign markets. These transport fees have remained relatively low compared to states that have closer ties to the West, and Gazprom is now buying natural gas from the obedient Central Asian region at near market prices. Remaining in Moscow's good graces has its rewards in high price purchases and low transport fees. Kazakhstan's relations with the West are not necessarily problematic; rather, this country has been very successful at keeping an impressive balance of good relations between the West, Russia, as well as China.[93] The latter country and Kazakhstan have been talking long-term relationship and a Kazakh-Chinese pipeline via the Central Asian-China pipeline expansion.[94] Yet the hypothesis that closer diplomatic ties to Russia for post-Soviet states will lead to lower prices on Russian natural gas fails to be falsified. We find the opposite trend in prices for states that have integrated or are attempting to integrate with the West.

The hypothesis that countries will pay more for natural gas if aligned more with the West also fails to be falsified. The assertion that closer diplomatic ties to the West will lead to higher prices for natural gas for countries of post-Soviet space, as evidenced in the previous chapter, finds statistical significance in this random effects panel regression analysis. The "Western ties" variable shows positive and statistical significance at the 95 percent confidence level with Russian gas prices. Georgia, Azerbaijan, and Ukraine have all broken off diplomatic relations with Russia at points in time of the dataset and sought protection via investment, military training and hardware, or diplomacy with the United States, thus all have sought cosier relations with the West as a result. This shows evidence that when relations are broken off, post-Soviet states look to the US and the West to help, and they pay more for Russian natural gas as a result.

Militarized disputes between Russia and states of the former Soviet Union do not have significant effects on the fluctuations in natural gas prices. This falsifies the second hypothesis of this chapter. The armed conflict between Russia and Georgia and the subsequent higher prices Georgia had to pay was not a result of the conflict itself but of Georgia cutting off diplomatic relations with Russia and recalling its ambassador after the conflict. As the "Western ties" and the "post-Soviet" variables are statistically significant, when diplomatic rows happen between Russia and post-Soviet states, these states seek diplomatic, economic, and even military refuge with the West. Russia will then raise the natural gas prices for these countries for this insubordination. Therefore, we have failed to falsify that Russia uses coercive energy policy in the post-Soviet space, and we have evidence that Russian and other states of the FSU's foreign policies affect the price of subsidized gas.

Moving on to the control variables, the "pipeline" variable is found to have negative statistical significance. This indicates that being a transport country for Russian gas or being endowed with one's own supply is a significant factor when deciding how much a country will pay for natural gas. Belarus, Ukraine, Azerbaijan, and Uzbekistan are examples. These states are also able to push back against Russian energy coercion by demanding lower gas prices for their populations in exchange for lower pipeline transit fees. Gas crises with Ukraine and Belarus, discussed in earlier sections of this chapter, have been Russia's response. The post-Soviet pipeline states' tactics of demanding lower prices have seemed to work, as evidenced by the results in Table 5.2. Russia is countering this leverage that pipeline states have used against it by offering lower prices for increased control over the post-Soviet state's pipeline infrastructure, which has been coined the assets-for-debts program.[95] This gradually takes the leverage away from these transit states and may result in more coercion via a subsidized pricing scheme when these states get out of line in Moscow's view.

Russia and Gazprom have also begun to take away the leverage of transit states by building pipelines that circumvent their territories altogether. This takes away any leverage transit states may have on Russia. The Nordstream pipeline is now operational and circumvents Belarusian and Polish territory, reducing Gazprom's reliance on those transit states. The South Stream pipeline, now postponed for the forseeable future, would have circumvented Ukrainian territory, which is Gazprom's most important transit state to European customers. Russian energy coercion is now evident in new forms.

The "ethnic Russian" variable is positive and statistically significant at the 95 percent confidence level. This is because of Moscow's perception of mistreatment by some post-Soviet states of these Russian ethnic minorities. Examples are replete: the ethnicity laws in Estonia in the 1990s, the bloody ethnic divide between eastern and western Ukraine, the frozen conflict between the central Georgian government and the ethnic Russian Abkhazians and Ossetians, and the status of the Russian-speaking Transdniesters in Moldova. Russia has increased the natural gas prices these governments pay in reaction to the domestic policies of these states over the status of ethnic Russians. Russia is trying to rein these states in with the power it has—the power of energy.

The final control variables that show statistical significance are the regional ones, where the European variable is dropped by the statistical software so that regions have a basis for comparison. Only the Caucasus coefficient is significant, where this region sees lower prices relative to the European region of post-Soviet space. This is not surprising due to Azerbaijan's own supply and the fact that Baku now supplies Georgia with most of Tbilisi's gas needs, at a discount price compared to what Russia was charging. Armenia is the Caucasus state that is not salient to Russia, as it is relatively Russia-friendly compared to other post-Soviet states and therefore continuously receives below-market, subsidized natural gas prices from Gazprom. Next we examine the results of the fixed effects panel regression, which shows the individual effects of each post-Soviet state when it comes to natural gas pricing, where who you are matters when Russia decides how much to charge.

Table 5.3 shows the results of the fixed effects panel regression model, which controls for the individual effects of each dyad on the independent variable, analyzes the impact of variables varying over time, and controls for the differing effects each characteristic of each dyad may have; error terms for each dyad are considered different and not correlated with other dyads. Each dyad is therefore treated as a dummy variable so we can see these individual effects. As with the random effects model, the fixed effects model in Table 5.3 shows negative statistical significance with the "post-Soviet ties" variable. States that stay in touch with Moscow diplomatically pay significantly less than states that break off relations with Russia and seek diplomatic, economic, or military shelter with the United States and the West. The "Western ties" variable loses its significance in this model. Finally, as shown in the previous model, the "Russian ties" and "MID" impact factors are not statistically significant.

Moving on to the dyadic dummy variables, Table 5.3 shows that it does matter who you are when it comes to Russian natural gas prices. The omitted dyad, Moldova-Russia, is the basis of comparison for the

Table 5.3 Russian economic statecraft with natural gas prices ($ per mcm) with dyadic dummies

	Coefficient	t-score
Russia ties	0.141	1.18
Post-Soviet ties	−0.485***	−3.76
Western ties	0.066	1.32
MID	0.214	1.61
Estonia	1.370***	5.38
Latvia	0.354	1.78
Lithuania	0.501**	2.68
Ukraine	−0.284*	−1.82
Belarus	0.172	0.64
Moldova	Omitted	−
Armenia	Dropped	−
Georgia	−0.050	−0.27
Azerbaijan	−1.195**	−3.03
Turkmenistan	−0.042	−0.14
Uzbekistan	−0.142	−0.47
Kazakhstan	0.202	0.71
Constant	−0.056	−0.43

$p < 0.001***, p < 0.01**, p < 0.05*$

Source of data: IMF (2013), UN Comtrade (2013), Henderson et al. (2013)
N = 1206

rest of the dyads in the table. Therefore, any significance is in relation to Moldovan prices, which serves as a basis of comparison. Armenia is dropped by the statistical software due to covariation with one of the other dyads. The Baltic states of Estonia and Lithuania all pay significantly more than Moldova, most likely because of their pro-Western orientation. Economic statecraft by Russia has been acutely felt by Estonia outside of natural gas with oil, even though it is under the economic and military protection of the EU and NATO, respectively. On 2 May 2007, Russia's state railway monopoly stopped all shipments of oil and coal to Estonia in the midst of political fallout between Estonia and Moscow over the relocation of a Soviet war memorial statue from a square in Tallinn.[96] Considering the series of cyber attacks that happened simultaneously (see Chapter 4), it seems that Moscow is easily offended, especially by those with a pro-Western orientation.

The Belarusian dyadic variable is not statistically significant. However, the Russia-Ukraine dyad shows negative statistical significance for this fixed effects model. Covered in detail earlier in this chapter, Ukraine's prices have fluctuated over the years, and this may be due to its importance to Moscow as the primary transport country. Furthermore, Ukraine has seen pro-Western as well as pro-Russian governments during this period, which could also explain the fluctuations in prices. For example, after pro-Western President Viktor Yushchenko took office in 2005, Gazprom demanded a sharp increase in the price of natural gas that Ukraine received.[97] Gazprom, and for all intents and purposes the Russian government, demanded an increase for its natural gas from $50 per mcm to $230 mcm, which was the current market price. Ukraine rejected this proposal, and Russia cut off natural gas supplies to Ukraine on 30 December 2006. Thus began the Russo-Ukrainian gas crisis of 2006, which came to an end when Gazprom and Kiev met halfway, with Ukraine paying more for gas and Gazprom agreeing to continue to pay transport fees, albeit at a discounted price. Also covered in detail in the earlier parts of this chapter, another example of coercive energy policy lies with Moldova, which has suffered higher energy prices even when making concessions to Moscow. Moldova's desire for Western integration has made compliance moot in the eyes of Gazprom and the Russian state.

The Russia-Georgia dyads show statistical significance. This is surprising due to the high level of attention and conflict this country has received from Russia and the international community. Something else besides energy concerns is driving the tensions with Georgia and Russia. Furthermore, Azerbaijan pays less than Moldova, primarily because it has its own supply of gas and has been able to diversify its supply outside of Russian, post-Soviet, and European markets. The gas Azerbaijan buys from Russia for transport south, therefore, is at a discount price where the Azerbaijanis have been able to coerce the Russians with energy—something not seen by any other post-Soviet state. Azerbaijan's increasing coziness to the United States means that it will be a thorn in Russia's side for the foreseeable future. Finally, the three Central Asian states in the model show statistical insignificance with Russian natural gas prices.

These panel regression analyses show that closer diplomatic ties with Russia will lead to lower natural gas prices for a former Soviet state, while breaking these diplomatic ties and seeking cozier relations with the West will be punished with higher natural gas prices. Militarized disputes are not explanatory factors for both the random effects model and the fixed effects model. Russia has yet to give up on regaining its former Soviet glory and will use the power it has at its disposal to punish any defecting post-Soviet states that stand in the way of Russia gaining the monopolistic political control in the region it sees as necessary to be viewed as a great power at home as well as by the global community.[98] As Russia is not the military and ideological superpower it once was, energy endowment has become a primary tool for post-Soviet Russian influence and the post-Cold War Russian "bully pulpit." With the completion of the Nordstream pipeline and the construction of the South Stream pipeline well under way, Russia will be able to circumvent Ukraine and Belarus, the most important Russian gas transport countries, effectively taking away any leverage these states may have on Moscow. This could bring Ukraine, which has been vying for pro-Western support, firmly under Moscow's grasp. Indeed, Russia has announced that it will begin to charge its former empire, as well as some domestic sectors, closer to market prices under the guise of meeting requirements for WTO membership.

Conclusion

Gazprom has a peculiar pricing system in post-Soviet space when it comes to natural gas. This is not because it gives subsidies to its domestic customers, something that is common in many energy-rich states. Pricing does not fluctuate due to selling off gas supplies to Europe and other foreign customers at market prices. Russia seems to punish with higher gas prices those countries of its former Soviet empire that get too close to the West. Having pipelines needed to transport Russian gas seems to be working for these post-Soviet states thus far, although Russia is being coercive in other aspects of energy policy, primarily through the construction of circumventing pipelines and the assets-for-debts programs, where Gazprom gives discounts to states that sell it majority shares in their pipeline infrastructures.

Russia has attained WTO membership status as of 2012, and its days of freedom to implement subsidized Near Abroad and domestic prices may be numbered. It has ignored the Energy Charter Treaty agreements of which it is a signatory, as the Putin Administration sees this agreement as one sided, where the West will come in and profit from Russian energy holdings, and Russia gets nothing in return.[99] Putin believes that Russians are fully capable of investing in and exploring for their own energy resources without Western help. Keeping the West out of Russia's affairs is one of Putin's top priorities, as demonstrated in the findings of this chapter.

Russia demonstrates its coercive power through the use of energy power politics against states with close ties to the West. The Nordstream and South Stream pipelines will ensure Gazprom will have a sizable stake in the EU market for years to come. These pipelines will also eliminate the dependence of Moscow on its former vassal states for transit, thus allowing Russia to retain the political dominance of the region. Gazprom's revenues continue to increase as a result, as these counties will have to pay market prices. If they do not, Moscow will cut off supplies, which it has done in the past, or force these gas-dependent states into making concessions beneficial to Gazprom. This could include the installation of pro-Russian puppet governments, the forced ascension to anti-Western alliances, and pro-Moscow votes in international organizations such as the UN. The question that remains is what is the outcome of these policies?

As the Moldova example shows, Russia's coercive energy policy results in the opposite of the policy's intended goals more times than not. Moldova has sold off its pipeline shares to Gazprom, agreed to pay more for gas than it can afford, lost its trade markets with Russia, allowed for Russian dominance of the Transinistria region, and has yet to succumb to economic and political dominance by Putin's government. Instead, Moldova has turned to the West. It has secured a much-needed grant for investment and is close to a free-trade agreement with the EU.

Ukrainian attempts at Western integration have also been met with a coercive energy policy from Russia in the form of mid-winter gas shut-offs. This brought a reluctant Ukraine to the negotiating table where transport and domestic gas prices were negotiated to Russia's advantage. The more recent 2009 shutoff occurred during the presidency of seemingly pro-Russian Yankuyvich; it seems that this coercive energy tactic has shifted the previously pro-Putin government towards more cooperation with the EU and the West, even in the midst of domestic turmoil where ethnic Russian separatists in the east are destabilizing the country. Therefore, these forms of energy coercion are having the opposite of the intended effects.

We infer from the findings of this chapter that the more Russia utilizes coercive energy diplomacy against a post-Soviet country, the more that state slips from Russia's sphere of influence and towards the influence of the West. Yet Russia continues to use these tactics, now more through pipeline diversions than over direct pricing schemes. States that break diplomatic closeness with Russia and move more towards the West are punished with tough pricing policies. States close to Russian influence are still punished and then move towards the West. Either course of action, as long as Russia uses coercive diplomacy in the energy sector, pushes states away from Russia. However, as Tsygankov proposes, Russia *needs* to view itself and have others view it as a great power.[100] As it does not have the military might it once had, it is using the power it does have, the power of energy, to dominate post-Soviet space and be satisfied with itself as a world power. The partially explained reasoning behind Russia's coercive energy policy in this chapter can perhaps be fully explained by Russia's identity issues. However, the next chapter shows a restrained Russia, a Russia that does not use coercion to its limits, and a Russia that is willing to work with its neighbors, friends, and traditional adversaries.

6
Energy Salience and Situational Coercive Diplomacy: Comparison of Coercive Energy Policy in the Caucasus and Central Asia

Introduction

Uzbekistan is a country of about 26 million people in the Central Asian region of post-Soviet space and is one of the most opaque and authoritarian states of the former Soviet Union.[1] When 23 local businessmen in the Uzbek city of Andijon of the Ferghana Valley, a region known for its disdain for the central government in Tashkent, were brought up on charges of Islamic extremism, 4,000 friends and relatives took to the streets beginning on 10 May 2005.[2] According to American defense lawyer Melissa Hooper, "This is more about (the businessmen) acquiring economic clout, and perhaps refusing to pay the local authorities, than about any religious beliefs."[3] Beginning on the night of 12 May, these peaceful protests turned violent when some of the protestors attacked and seized weapons from a military post and also stormed a prison, releasing some 4,000 prisoners onto the streets. On 13 May 2005, Uzbek President Islam Karimov ordered troops into the city. Punishment for this outbreak of civil disobedience was swift and harsh; many innocent bystanders were struck by government weapons in the attempt to quell the rebellion, with reported death tolls ranging from 34 to several hundred.[4] Known as "Bloody Friday," this incident brought the eyes of the United States upon the remote country with scorn and accusations of human rights violations.[5] In response, in July, Karimov ordered the expulsion of the American military from its airbase in Uzbekistan, which was being leased to the United States as an operations center for the War on Terror in Afghanistan.[6]

Russia's response to the bloody events in Andijon was quite different from that of the United States. Russian President Putin praised Karimov for his tough stance on Islamic extremism and reestablishing the rule of law.[7] "Sergey Ivanov even hinted at a connection to Chechen terrorists (an argument immediately picked up in Uzbekistan). While Putin and Karimov concluded that militants from Afghanistan had infiltrated Uzbekistan in a professional and 'thoroughly planned operation' (which was to remain their common line), the UN began its own investigation of the events."[8] Russia touted the concept of state sovereignty and self-determination in its defense of Karimov's actions in Andijon. It is not the West's business to criticize state actions within its own borders. These conflicting views on violent events in post-Soviet space between the West and Russia also came about three years later in August 2008 in the Caucasus state of Georgia.

When Georgia became an independent state in 1992, two regions, South Ossetia and Abkhazia, also sought independence from the government in Tbilisi or reintegration into the Russian state.[9] Abkhazians and Ossetians are tied to the Russian ethnicity and speak the Russian language. A bloody civil war ensued in Georgia in 1992–93, and Russia supplied the rebellious enclaves with military hardware and logistical support. The conflict ended with a stalemate, where Russian peacekeeping troops have remained ever since. Georgia is a small country of about 5 million that borders Russia in the north, the Black Sea to the west, and Turkey, Armenia, and Azerbaijan to the south. Home of Josef Stalin, this country in post-Soviet space has important historical ties to the Russian great power identity and has also been the biggest thorn in post-Soviet Russia's side in terms of ethnic violence and Western integration attempts in the former Soviet Union.[10]

This frozen conflict turned hot when pro-Western Rose Revolution champion President Mikhail Saakishvili made good on his campaign promise to reassert Georgian sovereignty within its own borders and sent troops into South Ossetia. Although it is still unclear who shot first, an interstate conflict ensued between Russian and separatist troops in both regions and Georgian regulars on 7 August.[11] The Georgians were heavily defeated, and hostilities ceased on 12 August, with Russia recognizing the independence of South Ossetia and Abkhazia and positioning more troops in the breakaway regions. This time, it was the Georgian government that was vilified with human rights violations and unprovoked violence in Russia and the United States/West crying foul of breaches of sovereignty and self-determination.[12]

Why the seemingly hypocritical responses by both Russia and the West over these violent clashes in two different regions of post-Soviet space? These security-related issues illustrate the different levels of salience Russia has for the regions of the Caucasus and Central Asia; this will be illustrated later in this chapter. The rivalry between the United States and Russia could be at play here, where each side is opposed to the other for no logical reason other than to oppose their perceived enemy.[13] The United States' heavy presence in the Caucasus when compared to Central Asia, the higher salience of the Caucasus region, the Russian public opinion on energy issues for each region, as well as Russian great power identity factors all contribute to the higher probability of energy coercion in the Caucasus over Central Asia by Gazprom and the Russian state. This chapter goes further and looks at the particular traits of the two regions that make one region more salient than the other and, through the lens of the theory of this volume, uncover why Russian energy coercion, in the form of the dependent variables of this chapter—natural gas prices, transit fees, and competitive/non-competitive pipeline projects—is apparent in the Caucasus and limited in Central Asia.

Western ties via international organization (IO) membership, democratization, transit routes, ethnicity, rivalry presence, public opinion, and identity measures, we argue, are all factors that lead down the path to Russian energy coercion in the Caucasus region, particularly with Georgia and Azerbaijan, or accommodation in Central Asia, specifically with Kazakhstan, Uzbekistan, and Turkmenistan. This chapter is structured as follows: First a research design section presents the hypotheses of this chapter and tailors them to the impact factors and dependent variables of the qualitative analyses; this is followed by the data presentation sections of the Caucasus and Central Asian regions and an assessment and concluding section.

Research Design: Energy Salience in Post-Soviet Space

Covered throughout this book, the region of the former Soviet Union is important for Russia's status as a great power, and therefore Russia has a special role and self-proclaimed privileges when it comes to the political and economic fate of the states that comprise its Near Abroad. Some regions, however, are more important than others.[14] In this chapter, we develop an energy salience scale and find that the European and Caucasus regions are relatively much more salient than the Central Asian region. Therefore, in this analysis of the latter two regions, we expect there to be non-market-based, more coercive natural gas prices and transit fees charged by Gazprom to the Caucasus states relative to those of Central Asia and for there to be competition between Russia and states outside the former Soviet over pipeline routes that go through as well as circumvent Russia. The Nabucco (Western- and American-backed) and South Stream (Russia- and Gazprom-backed) competing pipeline propositions through the Caucasus are examples of this competition and will be discussed in the following sections.

> H1: *If the energy salience level is high for a state and region of post-Soviet space, coercive energy policy will be used by Russia to keep the countries of this region within its sphere of influence.*
>
> H2: *If the energy salience level is low for a state and region of post-Soviet space, accommodative energy policy will be allowed by Russia, where Chinese energy investment will be tolerated.*

American geopolitical involvement has also been higher in the Caucasus region in relation to the Central Asian region. Furthermore, primary Chinese external investment and political involvement has been increasing in the past few years in Central Asia. One would expect that because of the salience of post-Soviet space to Russia and its assertion that the region is in its exclusive sphere of influence, that Russia would not react well to either American or Chinese incursions in the former USSR. However, because Russia is involved in a regional rivalry with the United States, and not China, we expect to see higher coercion in the Caucasus in relation to Central Asia, as the United States is the primary external investor for the states of the Caucasus. We will discuss the bilateral relations of Russia and each great power over each region to illustrate this point.

> H3: *States in post-Soviet space that allow American energy investment as its primary external funding source will have a higher likelihood of Russian energy coercion than those states that allow Chinese energy investment as their primary external funding source.*

Remembering our framework chapter, we hypothesize that public opinion can be an important measure for foreign policy actions by states. This is not to say that public opinion drives the actions by states, particularly the actions of semi-authoritarian states such as Russia. What public opinion can do is justify the actions that states take and allow their implementation to occur more smoothly. Leaders can be emboldened to make the decisions that they already wanted to make if the public has consensus on certain issues. We therefore expect that if the Russian public has consensus on the salient issues of the theory of this volume, the probability of coercive energy policy by the Russian state and Gazprom is higher. We expect enmity perception of the states of the Caucasus and more friendly perception for the Central Asian states. We present polls in the following sections taken by both Russian and Western pollsters.

> H4: *The closer to unanimous the public opinion on energy issues is, the more public opinion serves as a cue to the Russian governments as to how salient an issue is, and this motivates the Russian government to act coercively or seek cooperative outcomes.*

Finally, through the constructivist lens, we hypothesize that post-Soviet Russia is vying for a great power identity (Feklyunina 2008, 2012; Tsygankov 2006, 2010), not only with itself and its people but also with the international community.[15] Russian military power is a shell of its former Soviet self; therefore, using its natural gas and pipeline monopolization in the regions of the former Soviet Union as an instrument of power is all the more salient for the Russian foreign policy elite.[16] The following sections will also give a historical, cultural, and economic analysis of Russia's relations with each region that supports evidence for this peculiar need of a great power identity for Russia, which will also contribute to the explanation of the use of coercive energy tactics in the Caucasus and the lack thereof in Central Asia.

> H5: *The three regions of post-Soviet space—the European, Caucasus, and Central Asian—have different great power identity salience to Russia and its people, and the higher the identity salience for a region, the higher the use of coercive energy policy by the Russian state.*

Based on the four impact factors above, we expect the states of the Caucasus to pay non-market, coercive average natural gas price and pipeline transit fees more so than the Central Asian states, which are exporting countries and sell their gas to Gazprom at near market prices, from the years 2001–11. The unit of measurement for this dependent variable is US dollars per million cubic meters ($/mcm). We also expect there to be more pipeline competition in terms of Russia competing with outside powers for pipeline transit route projects, built and proposed, that circumvent Russia or go through its territory, in the Caucasus region in relation to Central Asia. A tabular presentation of these projects will be in the dependent variable sections of this chapter. Next we present the data of the impact factors as well as the dependent variables—gas pricing and pipeline competition—to uncover why Russia uses coercive energy tactics more in one region of the former Soviet Union in relation to another.

Here we use Hensel's territorial issue salience measure to find out how important the states and different regions of the former Soviet Union are to Russia when it comes to issues over natural gas.[17] This concept and measure is adjusted for Russia's salience scores for energy in post-Soviet space. The question we answer in this section is how salient is the energy issue to Russia for each post-Soviet state as well as each of the three regions of the former Soviet Union? Russia is vying for territorial claims for the monopolization of transport and pricing of natural gas to European countries, regional political hegemony, and protection of ethnic Russians living in the countries of the former Soviet Union. A measurement of issue salience is tricky, yet scholars have attempted and succeeded in giving us guidelines on how to measure the importance of issues to particular states. Hensel's scale for the salience of territorial issues is derived from the Issue Correlates of War (ICOW) dataset on territorial disputes.[18] The data for this set is limited to the Americas and therefore is not available for disputes over the post-Soviet region. We are able to adjust this scale through an examination of the issues that encompass the stakes of this all-important region for Russia, as well as for each possible dyad between Russia and the other 14 states of the FSU. How salient is the energy issue to Russia with each state, and is it willing to use coercive energy tactics or even fight for what it perceives to be rightfully its own?

The issue salience scale developed by Hensel is a 12-point index where a point is given to each state if it meets each of the six salience indicators.[19] Therefore, an issue is considered very salient if the dyad gets a score of 12 and not very salient if the dyad gets a score of one. An issue is considered highly salient if the score lies between eight and 12, moderately salient if the score is between five and seven, and of low salience if the score is four or less. The six factors measuring issue salience for territorial claims are as follows[20]:

1 The presence of a permanent population.
2 The (confirmed or believed) existence of valuable resources.
3 The strategic economic or military value of the territory's location.
4 The existence of a state's ethnic and religious kinsmen in the territory.
5 Whether or not the territory is considered part of the homeland or is a dependency.
6 Whether or not the territory is part of the mainland or is offshore.

Along with these initial salience indicators, Hensel incorporates other factors contributing to either peaceful settlements or escalating conflict.[21] They are issue management, past interactions, institutional context, and the characteristics of the "adversaries." Therefore, eight points are added to the scale to give it a 20-point index. These supplemental factors are only given scores of "2" or "0" and are operationalized according to this scale.

7 Issue management: Entails how issues and their stakes have been handled in the past—either peacefully or with force.

A score of "2" is added if past interactions involved force, "0" otherwise.

8 Past interactions: Only pertain to past military conflict between states for the past 15 years.

This measure counts the number of militarized interstate disputes (MIDs) for dyads occurring within the 15-year threshold.[22] Scores of "2" are given to the pair of states if there have been recent MIDs, "0" otherwise.

9 Institutional context: Measured as a count of multilateral treaties and institutions calling for the peaceful settlement of disputes that both states have signed and ratified."[23]

Hensel looks at the presence of signed treaties or membership in institutions and codes a "2" for their presence, "0" otherwise.

10 Adversaries: Measures whether or not the pair of states under analysis are democratic, mixed between democratic and non-democratic, or both non-democratic.

The democratic peace research paradigm has empirically found that democracies rarely resort to armed conflict over disagreements.[24] A score of "2" is given if there is the presence of a mixed dyad (democratic and non-democratic), "0" if it is an authoritarian pair of states or democratic pair of states.

We alter this scale for the purposes of this specific analysis of energy coercion and not solely territorial disputes. This scale is meant to measure salience for Russian policymakers. This salience scale is a mixture of being intrinsic—where change is difficult and can only happen with natural gas discoveries, pipeline construction, regime change, or huge demographical shifts—and relational—where changes in international organization membership could quickly change the dynamics of relations between Russia and post-Soviet states. The first five energy salience scores are the more intrinsic ones, while the sixth one is the more dynamic and relational one. The altered coercive energy salience indicators are part of a 13-point index and are as follows:

1 Natural gas: Whether or not the post-Soviet state has natural gas endowments of its own and therefore can use Russian pipelines to transport its gas or alternatively use a non-Russian pipeline export route.
2 Transit: The post-Soviet state is a current transport state for Russia and serves as a go-between to markets outside of post-Soviet space, or it is a transit country that circumvents Russian pipelines and also serves external markets.
3 Ethnic: Whether or not a post-Soviet state has 15 percent or more ethnic Russians residing in its territory (15 percent is the average amount of Russians who live in all countries of the former USSR; therefore, countries that have more than 15 percent are above average, and this serves as a logical cut point). The work of Ayres, Saideman, Jenne, Saideman and Ayres, and Jenne, Saideman, and Lowe find that the amount of external support that ethnic minorities receive in the face of discrimination has important international and regional stability implications.[25] Here we argue that the more ethnic Russians in a post-Soviet space, the higher the probability for radicalization and therefore the higher the probability for a crackdown on these ethnic Russian minorities. This will lead to a reaction by Russia in the form of coercive energy policy. Therefore, when a state has over 15 percent of ethnic Russians comprising its population, it is more likely that Russia will monitor the treatment of these people and react with coercive energy policy if they are perceived as being mistreated.
4 Past interactions: Looks at military conflict between states for the past 15 years. This measure counts the number of militarized interstate disputes (MIDs) for dyads occurring within a 15-year threshold.[26]
5 Adversaries: Estonia is democratic, Kazakhstan is authoritarian, while Russia is not considered a full-fledged democracy and also is considered an authoritarian regime.[27] Furthermore, as Kramer notes: "Putin (with his own authoritarian bent) seemed most comfortable when dealing with authoritarian leaders who will support Russia's interests and align their countries squarely with the CIS."[28] Therefore, any mixed dyads (democratic-authoritarian) involving Russia implies ties to the West, which is an important measure for this chapter. Western ties lead to discord and thus raise the probability of Russian coercive energy policy. All of the above measures receive a "2" if the variable is present, "0" otherwise. The measures for each post-Soviet state are listed in Table 6.1, followed by the regional salience scores in Table 6.2.

Table 6.1 Natural gas salience scores between Russia and former Soviet states (Low 0–4, Moderate 5–9, High 10–13)

Post-Soviet State	Natural Gas	Transit	Ethnic	Past Interactions	Adversary	Institutional	Total
Ukraine	0	2	2	2	2	2	10
Georgia	0	2	0	2	2	2	8
Azerbaijan	2	2	0	2	0	2	8
Latvia	0	0	2	2	2	2	8
Moldova	0	0	0	2	2	2	6
Estonia	0	0	2	0	2	2	6
Lithuania	0	0	0	2	2	2	6
Kazakhstan	2	2	2	0	0	0	6
Belarus	0	2	2	0	0	0	4
Turkmen.	2	2	0	0	0	0	4
Uzbekistan	2	2	0	0	0	0	4
Kyrgyzstan	0	0	0	0	2	0	2
Armenia	0	0	0	0	2	0	2
Tajikistan	0	0	0	0	0	0	0

Table 6.2 Former USSR regional sub-complex salience scores

Regional Sub-Complex	Salience Score
European	6.67
Caucasian	6.00
Central Asian	3.20

Also in Table 6.1 is the last and most dynamic part of our altered coercive energy salience measurement scale:

6 The institutional context: Here we measure the level of Western or Moscow-based ties more specifically and more dynamically. We use three Western-based organizations and give each post-Soviet state a score of "2" when membership is present for each organization. The three Western institutions are the European Union (EU): either the state is a full member, or a member of the EU's (economic) Neighborhood Policy, where reduced barriers and economic integration are encouraged; the North Atlantic Treaty Organization (NATO), a traditionally adversarial organization to Russia; and GUAM, the acronym for the four countries who have signed an agreement to cooperate more with the West: Georgia, Ukraine, Azerbaijan, and Moldova. A score of "–1" is given to states who belong to the more Russia/Eurasian-friendly organizations of the Collective Security Treaty Organization (CSTO), post-Soviet Russia's answer to NATO; the

Shanghai Cooperation Organization (SCO), the Russo-Chinese-led economic organization in post-Soviet space; and the Eurasian Economic Community, post-Soviet Russia's answer to the EU. The logic behind the "−1" score is that when an FSU state is a member of one of these organizations, any salience score in other areas pertaining to energy in post-Soviet space should be mitigated and nullified. However, if a post-Soviet state belongs only to Russian-friendly organizations, that state will receive a "0" in this category. We now have a scale where the maximum score is 13 and the lowest score is 0.[29] We expect dyadic scores of 0 to 4 to be of low salience for Russia, scores of 5 to 9 to be of moderate salience, and scores of 10 to 13 to be of high salience for Russia, and therefore these states should expect Russian coercive energy policy. All scores are compiled looking at news reports, scholarly articles, organizational websites, and relevant datasets.[30]

Next we compare energy salience of the Caucasus and Central Asian regions.

Caucasus Energy Salience: Ripe for Coercive Energy Policy

The Caucasus region holds an important place in the Russian great power identity due to the historical high stakes in security issues and more recently the region's important strategic economic implications. According to Nygren, "Russia and the three former Soviet republics of Georgia, Azerbaijan, and Armenia are so strongly interlinked as to be almost oversensitive to changes in any one of the other relationships in the sub-complex as well as to changes in general power balances. As part of Russia's own regional security complex, it is indeed the most unstable and therefore most volatile of the sub-complexes."[31]

The next section will present findings on the salience of the Caucasus region to the Russian energy policy elite, which is further exacerbated by the heavy political and economic presence of Russia's longtime rival, the United States, backed by the public opinion polls of the Russian people on various issues of this region, and how the Caucasus is important to the reassertion of Russia's great power identity, which was perceivably lost with the fall of the Soviet Union.

Salience of the Caucasus

Tables 6.1 and 6.2 give a summary of the energy (natural gas) salience scores for the three states of the Caucasus as well as the overall average energy salience score for the region. As far as its foreign energy policy pertaining to natural gas and natural gas pipelines for Russia and these countries, Georgia and Azerbaijan fall in the high-salience category, with Armenia in the low-salience category. Georgia is highly salient to the Russian state mainly because of its positioning as the transit country that circumvents Russian pipelines, its defiance of Russia's wishes to fulfill its great power identity as the dominant state of the Caucasus and its numerous attempts at Western assimilation. It is a member of GUAM (which stands for the members Georgia, Ukraine, Azerbaijan, and Moldova), the organization funded and backed by the United States that promotes the development of democracy, stable economic development, regional and international security, and fast-tracking integration within the European security and economic sphere of influence.[32] This organization is seen as a US-backed counter to the Russian-dominated CIS, an organization whose future looks bleak as post-Soviet members such as Georgia have left it. Membership in GUAM, therefore, is something that the Russian government frowns upon; therefore, coercive energy policy is expected with its members.

Georgia is also a member of the European Union's Economic Neighborhood Policy, which is very similar to the US-backed GUAM as it has "the objective of avoiding the emergence of new dividing lines between the enlarged EU and our neighbors and instead strengthening the prosperity, stability and security of all. It is based on the values of democracy, rule of law and respect of human rights."[33] Although perhaps not as disconcerting to the Russian energy policy elite as membership in GUAM, this membership still denotes Western integration away from Moscow's political and economic grasp in its perceived exclusive sphere of influence, which will lead to coercive energy policy on the Georgian state.

Georgia is also a developing democracy, which implies Western integration. Touched upon in the previous section, Georgia under Saakishvili has been close to the United States in terms of security policy. Georgian troops were sent to Iraq with President George W. Bush's coalition of the willing as part of the superpower's larger War on Terror. The United States has rewarded Georgian allegiance with economic aid, military hardware and training, and heavy investment in oil and natural gas pipelines circumventing Russia (BTC, BTE, and the proposed Nabucco).

There have, however, been drawbacks to this American coziness in the form of military action by Russia in 2008. Perhaps emboldened and overconfident that he had the full military support of the United States, President Sakkishvili may have acted in haste to reassert Tbilisi's control over the breakaway regions of Abkhazia and South Ossetia. He paid the price with an overwhelming defeat by the Russians and the temporary shutdown of the BTC and BTE pipelines, the Russia-circumventing pipelines that feed the energy needs of important EU customers. Kramer points out that this could have been a secondary motive for the five-day conflict between Russia and Georgia.[34] Russia is trying to paint the picture that Georgia is an unstable country that not only does not have control over its territory but also is an unreliable transit country for European oil and gas customers, with the only alternative transit routes coming from this region traveling through Russia. Therefore, Russia has important geopolitical and geoeconomic reasons for keeping Georgia within its sphere of influence, and punishment for this disobedience has been enforced not only with armed conflict but coercive energy tactics.

Azerbaijan, the only Caucasus state that has significant reserves of its own natural gas and oil, is quite salient not only to Russian but also Western energy companies. Furthermore, the state has also been involved in an interstate war in which Russia has sided with Armenia, Azerbaijan's adversary of the Nagorno-Karabakh disputed region. This conflict has displaced some 600,000 persons and has killed about 30,000.[35] The primarily Christian Orthodox Armenia has seen Russian support more than the Islamic Azerbaijan. During the war that lasted from 1988 to 1994, it was the Soviet Union and later Russia that sided with Armenia in this yet-to-be-settled dispute at the expense of Azerbaijani sovereignty and self-determination. Yet due to the energy importance of Azerbaijan, Russia has yet to move militarily on the side of Armenia to end this bloody and frozen conflict.[36] As Armenia has depended upon Russia for its livelihood in terms of military, diplomatic, as well as energy support, Western integration for the Caucasus state has been limited, and Western governments and energy companies have not courted Armenia and their territory for pipeline transit access. Georgia has been the main beneficiary of the Armenian orbit around Moscow over the West. As this conflict remains frozen, Armenia remains close to Russia while Azerbaijan is slowly choosing Western security and economic integration. This keeps the salience level high and, because Azerbaijan has its own resources, it more likely that Russia will coerce the government in Baku with competition with the Western investors of pipeline routes that attempt to circumvent Russian territory.

Azerbaijan is an opaque and autocratic state, which somewhat mitigates its salience score with Russia presented in Table 6.1; however, it is a member of the US-backed GUAM as well as part of the EU's Neighborhood Policy.[37] Its membership in GUAM has brought the heavy investment of the American government and American energy companies for exploration of new oil and gas fields and their extraction, for the BTC and BTE pipelines, freeing the state from Russian monopolization and political dominance. Its membership in the EU Neighborhood Policy has also made Azerbaijan a favorite supplier of oil and natural gas to the European market, especially after the gas shutdowns by Gazprom in Ukraine in the winters of 2006 and 2009. These crises in the European region of post-Soviet space are discussed in the next chapter. This increased Western presence in Azerbaijan, therefore, has kept its salience score with Russia in the high category. Exacerbating the Russian foreign energy policy elite even more is the fact that its longtime rival, the United States, is the country gaining influence in the Caucasus at the expense of Moscow.

Therefore, heavy American investment for energy projects in any region of post-Soviet space should be met with Russian energy coercion. Remembering the previous section, Azerbaijan is also quite salient to the Russian energy policy elite in that it is the only Caucasian state that has significant reserves of oil and natural gas. Baku, the capital city and port on the Caspian Sea, is at a strategically important energy pipeline crossroads and is important to Gazprom, Rosneft, and Transneft for Russia and energy conglomerates such as Exxon-Mobil, Royal Dutch Shell, and British Petroleum for the West. Furthermore, Western, especially American, investment in the energy infrastructure of Azerbaijan has been high.[38] Looking at Table 6.3, since 2005, only neighboring Turkey has invested more than the US in Azerbaijan's energy, and the UK, Germany, and Russia, respectively, round out the top five.[39] However, since 2009, the United States has been the most heavily invested country in Azerbaijan. This is supporting evidence for the second hypothesis of this chapter. However, there must be evidence that Russia is in fact being coercive with Azerbaijan, and these results are presented in the dependent variable section of this Caucasus analysis. Azerbaijan, therefore, is arguably the most geopolitically and geoeconomically important state in post-Soviet space in terms of competition over natural gas transport to Western markets.

Table 6.3 Top five FDI energy investors in Azerbaijan ($/million)

	2005	2006	2007	2008	2009
USA	$24.8	70	78	108.8	117.6
Turkey	96.2	136.6	109.2	60.8	76.8
Russia	5.1	4.6	10.7	5.8	50.3
Germany	21.5	17.4	22.9	48.2	38.8
Iran	1.2	17.5	4.6	-	6.8

Source of data: US Embassy in Baku (2013): http://azerbaijan.usembassy.gov/economic-data.html

Public Opinion of the Caucasus States and the United States

A major tenet of the theory of situational coercive diplomacy is that public opinion can be a reliable supplementary predictor of foreign policy actions by states. By no means are we asserting that public opinion is a cause for or drives foreign policy actions, especially in a state such as Russia, where the media is largely state owned and censored. However, it can help push foreign policymakers to act coercively, as Russia does with its coercive energy policy in post-Soviet space.[40] This section tests the public opinion hypothesis for the theory of coercive energy policy. Below are Russian public opinion polls about Caucasus states from the Russian-based Levada Center and the Russian Public Opinion Research Center (VCIOM) as well as from the US-based Pew Research Center's Global Attitudes Project.[41]

Table 6.4 shows evidence for Russian domestic sources of rivalry perpetuation.[42] For polls taken of Russians from 2005–11, when asked to name five states that are unfriendly or hostile to Russia, three states in the Caucasus appear. Georgia and, although not a geographical part of the region still an important political player in its fate, the United States, are the two most adversarial states in the entire poll. These are also the two states that are currently rivals of Russia.[43] Therefore, when the only two rivals of Russia are present in one region of post-Soviet space, coercive energy tactics are almost certain to be utilized by the natural gas arm of the Russian state, Gazprom. A small percentage of Russians also put Azerbaijan in their top five for all six years, which increases the chances of more coercion, especially with the heavy American investment in the energy- and pipeline-rich Caucasus state.

The Russian public's blame for the cause of the August 2008 conflict between Georgia and Russia is quite telling of the continuing animosity towards the United States and its motives in the Caucasus region. Table 6.5 shows that half of the Russians polled in September 2008,

Table 6.4 Russian opinion of states it considers unfriendly or hostile in the caucasus

Respondents asked: Name five countries that you would consider unfriendly or hostile to Russia.

	2005	2006	2007	2009	2010	2011
Georgia	38%	44	46	62	57	50
United States	23	37	35	45	26	33
Azerbaijan	5	4	4	2	3	5

Source of data: http://en.d7154.agava.net/sites/en.d7154.agava.net/files/Levada2011Eng.pdf

one month after the conflict, see the United States as having motives to spread its influence in the strategically important Caucasus region, beginning with Georgia. With 32 percent of Russians of the opinion that ethnic discrimination was the motive behind the conflict, this leaves only 5 percent of Russians asserting that their own government was at fault fault for the initiation. Such a high consensus of the Russian public citing American or Georgian belligerency for the conflict involving Russia, according to the theory of coercive energy policy, will lead to the certainty of Russian energy belligerency against Georgia as well. Georgian natural gas prices and pipeline fees should be relatively higher than other states of post-Soviet space, and competitive pipeline tactics should also be present in the small Caucasus state.

Table 6.5 Russian opinion of the cause of the August 2008 Russian-Georgian conflict

Respondents asked: In your opinion, what is the ultimate cause of the ongoing conflict in South Ossetia?

Georgian government pursuing a policy of discrimination against the Ossetian people	32%
Authorities of unrecognized Ossetia and Abkhazia trying to hold on to power	5
Russian officials conducting divide-and-rule policy to maintain authority in the Caucasus	5
United States seeking to spread its influence in the Caucasus region	49
No opinion	10

Source of data: http://en.d7154.agava.net/sites/en.d7154.agava.net/files/Levada2009Eng.pdf. September 2008.

Table 6.6 Russian opinion of Western support of Georgia in the August 2008 conflict

Respondents asked: In your opinion, why did governments of the West support Georgia in the South Ossetia conflict?

	2008	2009
Because Russia infringed upon Georgian territorial sovereignty	7%	6
Because Russian shelling killed civilians	8	10
Because Russian actions sparked conflict in other regions, specifically Abkhazia	5	5
Because they want to impair Russia and force it out of the Transcaucasus	66	62
No opinion	14	17

Source of data: http://en.d7154.agava.net/sites/en.d7154.agava.net/files/Levada2009Eng.pdf.

Table 6.6 goes further and asks Russians why there was widespread Georgian support in the West, with Russia seen as the belligerent and aggressive state in the 2008 conflict. The overwhelming majority opinion on this issue is telling: Nearly two-thirds of Russians in both 2008 and 2009 see the reasons behind widespread Western support through an imperialist zero-sum competition with Russia. Russians see the Georgians as a tool of the West whose main motive is to push Russia out of its exclusive sphere of influence in the Caucasus. Russia, therefore, acted militarily to defend its honor and great power identity. An alternative explanation for Russian motivation for the conflict is that Russia wished to discredit Georgia and its role as a non-Russian gas transporter to the European market and make it seem unreliable.[44] Russia saw its great power identity and its energy monopolization slipping in the Caucasus; thus the conflict was a tool to reassert both the former and the latter. Great power identity and dominance over the former Soviet Union is also achieved through coercive energy tactics, which is the final impact factor contributing to the motives behind this peculiar Russian policy.

Russian Great Power Identity in the Caucasus: This Land Is Our Land

The last impact factor that leads to coercive energy policy in post-Soviet space, specifically the Caucasus, is the region's historical importance to modern Russia's great power identity, both to its government and people, but also how it is viewed by other governments and people around the world.[45] According to Tsygankov, "Russia is returning to its identity as a regional great power. Its priorities once again include security and prosperity in the territories adjacent to its borders, and it increasingly sees itself as a European power with special relations to Asia and the Far East."[46] Therefore, for Russia to be identified both at home and abroad as a great power, it must rebuild its economy via energy and secure its borders where its influence in post-Soviet space trumps any other influence attempts by any other great powers, such as the United States or EU.

The Caucasus is the most volatile regional complex of Russia's Near Abroad and contains the state also seen as the biggest regional thorn in its side, Georgia, supported by its global and historical thorn in its side, the United States. Furthermore, during the brief 2008 conflict, Azerbaijan backed Georgia's territorial integrity and allowed the entrance of Georgian refugees.[47] It also began more heavy diplomatic courtship of the United States and its protection of its pipelines circumventing Russia.[48] This has raised the salience of issues in this region for Russia, and some go as far as to say that these heightened Western incursions in the

Caucasus along with Georgia's overt declarations about joining NATO are what pushed Russia over the edge and justified its use of armed force in South Ossetia and Abkhazia in August 2008.[49]

The historical ties for Russia to the Caucasus go back centuries, and the region was once the convergence point of three great empires: the Ottoman, Persian, and Russian. Russia sees this region as a place of Russian honor, courage, and sacrifice. Russia wrested the Azerbaijani clans from the oppressive rule of the Persians in 1812. It annexed the Georgian kingdom in 1801 to keep it safe from the Ottomans, and it absorbed Armenia after World War I in the face of Armenian genocide at the hand of the Turks. Furthermore, Georgia is the home of the infamous Soviet leader Josef Stalin, who is still getting mixed reviews by the Russian public as to whether or not he is to be remembered as a tyrant or a great leader who modernized Russia and made it a superpower. Regardless, this historical region under the political, economic, and military influence of the longtime rival of Russia, the United States, does not sit well with the Russian foreign policy elite and, therefore, the chances of coercive energy policy in the region is higher here than in Central Asia.

According to Feklyunina, "Identity is seen as one of the key concepts in explaining why a country takes particular actions in the international arena," and "one of the basic needs of the state that shapes its vision of national interests is, from the constructivist perspective, the need for collective self-esteem."[50] Tysgankov and Tarver-Wahlquist go further and assert that Russia's national honor has suffered since the Soviet breakup, and regaining this great power identity and restoring its honor is at the top of Vladimir Putin's foreign policy objectives.[51] Therefore, getting the United States out of the Caucasus and bringing troubled states such as Georgia and Azerbaijan back into its grasp, either by cooptation or coercion, is something that Russia is able and quite willing to do.

Feklyunina explains the reasons behind Russian coercive energy policy in the Caucasus best from the constructivist perspective when she notes that the "sharp contrast between the perceived strength of the Soviet Union and the Russian Empire and the obvious weakness of the Russian Federation led to growing dissatisfaction with Russia's role in the international arena. This dissatisfaction was shared not only by political elites, but also by many ordinary Russians, who welcomed a more active, even aggressive foreign policy aimed at defending Russia's national interests and positioning Russia as an equal partner of Western countries," and "for Putin's Russia it is not military but rather economic strength that is positioned as important. This shift can be explained by the fact that due to numerous economic and social problems it would be extremely difficult for the Russian army to pose a serious threat to Western countries."[52]

With the four major impact factors of Russian coercive energy policy in the Caucasus region predicting high coercion, next we look at the dependent variables for the Caucasus. We expect relative natural gas prices and pipeline transit fees in the Caucasus to be higher for Georgia and Azerbaijan, the states that rank high in the impact factors of the theory, when compared to the Caucasus state of Armenia, a state that has remained loyal to Russia. We also expect a high level of competition over pipeline routes to Europe between Gazprom/Russia and Western/non-Russian investors.

Caucasus Dependent Variables: High Prices and Pipeline Politics

Table 6.7 shows the natural gas pricing in dollars per million cubic meters ($/mcm) between the Caucasus and Russia from 2005–11, which includes the discount for each state's transit subsidy, if applicable.[53] There is evidence of natural gas coercion by Russia with Azerbaijan and Georgia. For Georgia, as part of the trade dispute with Russia over the issues of Abkhazia and South Ossetia, Russia increased the price of natural gas in 2006 to three times the price charged in 2005. This was in response to Saakishvili's increased hardliner rhetoric about regaining Georgian sovereignty and establishing political control over the separatist enclaves by any means necessary, including military force. Russia responded to this by boycotting Georgian wine, of which nearly most of the exports of this commodity went to Russia, as well as the steep hike in natural gas prices. Furthermore, a series of mysterious blasts crippled the pipelines from Russia to Georgia in the midst of one of the coldest winters in record.[54] In response, Georgia turned to its neighbor to

Table 6.7 Natural gas pricing in the Caucasus region ($ per mcm)

Caucasus State	Import/Export Scheme	2005	2006	2007	2008	2009	2010	2011
Azerbaijan	Import from Russia	45	105	235	-	-	-	-
	Export to Russia	-	-	-	-	-	244	288
	Export to Georgia	-	120	130	140	187	161	170
Georgia	Import from Russia	73	235	235	235	-	-	-
	Import from Azerbaijan	-	120	130	140	187	161	170
Armenia	Import from Russia	67	70	85	100	154	180	180

Source of data: Henderson et al. (2012).

the south, Azerbaijan, to fulfill its natural gas needs and began weaning off the dependency from Russia, with over 90 percent of its natural gas coming from its Caucasus neighbor by 2008. What is fuzzy is whether or not this was in any way correlated with the guns of August conflict, which was waged in the same year. What is clear is that Russo-Georgian relations have soured since this energy dispute, and even after the short conflict in 2008, Georgia has sought refuge in the Western security and economic umbrella, indicating that these coercive tactics by the Russian state are not changing the behavior of states to their liking.

Azerbaijan has its own reserves of natural gas and is therefore a key competitor to Russian energy hegemony in the Caucasus, and it can also act as an alternative transport state for Western energy customers and energy companies. Azerbaijani defiance towards Russia via competition both for natural gas and pipeline routes has made the energy issues salient for the Russian energy elite, as indicated earlier in this chapter and in the previous chapter. This has attracted Western investors, including those from the US government and private sector. This intrusion into post-Soviet space by the United States, Russia's rival, has led to coercive tactics against Azerbaijan. Table 6.7 shows the natural gas pricing interactions between Russia and Azerbaijan. Azerbaijan used to import Russian gas that was headed to Armenia via Iran at a very subsidized price in 2005, but as Azerbaijan learned of where this gas was going, its principal enemy Armenia, prices went up and the import of Russian gas ceased in 2008. Also contributing to this cessation was Azerbaijan filling the gap in Georgian natural gas at below-market prices as a result of souring relations between Moscow and Tbilisi, as well as the opening of the Russian territory-circumventing South Caucasus (BTE) pipeline in 2008. Natural gas trade all but ceased during 2008 and 2009, with Gazprom coming back to Azerbaijan and buying Azerbaijani gas at near market prices in 2010 in order to fill orders for the increasing demand from Europe. Therefore, Azerbaijan has been able to use its geopolitical and geoeconomic position, as well as its own energy reserves, to its advantage and keeps its distance from Russian dominance over its energy policy, which has drawn it closer to the American and Western sphere of influence, much to Moscow's chagrin.

Evidence for Russian energy coercion, or in this case the lack thereof, on Armenia is also evident in Table 6.7. Armenia has no natural gas reserves of its own and pays well below market prices, less than its neighbor to the north, Georgia. This can be attributed largely to Armenia's remaining friendship with Russia. Russia has backed Armenia in the Nagorno-Karabakh conflict with Azerbaijan, and for this Armenia has supported Russian foreign policy actions. Furthermore, Armenia is part of the Russian-led security organization CSTO. This has kept Russian gas flowing to Armenia at a discount rate, much lower than for Georgia, even though the gas must navigate through Georgia to Armenia via the North-South pipeline. For Armenia, it pays to be close to Russia.

Looking at Table 6.8, the competition between Russian and non-Russian-based pipeline routes in the Caucasus region is evident. With the opening of the South Caucasus (BTE) pipeline in 2008, the Russian natural gas transport monopoly has been broken. Azerbaijan, with the help of Western, primarily American, investment, can now deliver its natural gas to European customers without the dependence of the Russian-based and Gazprom-owned Baku-Mozdok and Blue Stream pipelines. Azerbaijan also has the option of exporting to Iran, although as of late it has ceased all natural gas trade as part of the American-led embargo on Iran, due to its questionable nuclear program. What is important to note is that the BTE pipeline is new, Western backed, and breaks Russia's transport monopoly to the West, and the competition for more transit routes between Russian and Western investors is growing, as evidenced in Table 6.9.

Table 6.9 shows that the competition is heating up in the Caucasus region over pipeline routes to the all-important European market. Three Russian-circumventing pipeline proposals, all with financing from Western-based companies and governments, have been proposed and seriously considered in the past five years. Especially in light of the 2008 Russian-Georgian conflict, these proposed pipelines have been increasingly questionable due to the unknown status of stability in the region, especially Georgia's control over its own territory. All three proposed Western-backed natural gas pipelines must travel through Georgian territory. This is evidence for the assertion that the 2008 conflict was not over the sovereignty and protection of ethnic Russians. This glaring energy issue could be a main contributing factor behind Moscow's decision to resort to armed conflict in Georgia, a conflict widely condemned

Table 6.8 Existing Caucasus natural gas pipeline

Pipeline	States on Route
South Caucasus (BTE)	Azerbaijan-Georgia-Turkey to Europe
North-South Caucasus	**Russia**-Georgia-Armenia to Armenia
Baku-Mozdok	Azerbaijan-**Russia** to Europe
Blue Stream (BS)	**Russia**-Turkey to Europe
Iran-Armenian	Iran-Armenia to Armenia
Kazi Magomed–Astara–Abadan	Azerbaijan-Iran to Iran

Source of data: Dellecker and Gomart (2011).

Table 6.9 Proposed Caucasus natural gas pipelines

Pipeline	States on Route
White Stream	Azerbaijan-Georgia-Romania/Ukraine to Europe
South Caspian	Turkmenistan/Central Asia to BTE to Europe
Nabucco	Azerbaijan (BTE) to Bulgaria-Romania-Hungary-Austria
South Stream	**Russia** to Bulgaria-Serbia-Slovenia-Italy

Source of data: Dellecker and Gomart (2011).

not only by the US and the West but by the larger international community. What is even more alarming is that the outcome of the war, if this assertion is true, may have worked. The only pipeline that has broken ground and started construction at the time of this writing is the Gazprom-backed South Stream pipeline. Many Western backers have shown less confidence and pulled their money out of the White Stream, South Caspian, and Nabucco pipelines, and these projects have been shelved for now. This is a Russian victory in the sense that it will have more control over the pipeline scenario in the Caucasus once South Stream is completed, and it took an armed conflict to get this control. This was Russian energy coercion at its most extreme, as military intervention has led to a favorable Russian energy policy. What might this mean for future endeavors where Russia seeks energy control in not only the Caucasus but other regions where Russia sees itself as having privileged interests?

This section has shown that the high salience of the Caucasus region; the high presence of Russia's principal rival, the United States; the public opinion results about the region; and the importance of the region to Russia's great power identity have all contributed to the high level of coercion of the Russian state on energy issues, salience that has even led to armed conflict in post-Soviet space. With this in mind, we now move to another region of post-Soviet space, Central Asia, to see if energy coercion is present where salience is low relative to the Caucasus, the US-Russia rivalry is relatively absent, public opinion does not see the region as something for concern to Russia's interests, and the importance of the region to Russia's great power identity is not as strong.

Central Asian Energy Salience: Limited Coercion

Central Asia has moved up on the security policy priority list of the United States and Russia since the invasion of Afghanistan in 2001; however, with these conflicts winding down and the US military presence dissipating, it is becoming a region of competition between Russia and another major power, China. The region includes Kazakhstan, Kyrgyzstan, Tajikistan, Turkmenistan, and Uzbekistan—all former Soviet states. Central Asia is important for its energy potential and transit role. Kazakhstan, Uzbekistan, and Turkmenistan are the three states with large reserves of natural gas, and this section will only focus on the relationship between Russia and these countries. These states are all landlocked and need the assistance of other countries and their territories to get their gas to external markets. They have three options: through Russia, through Azerbaijan and the West, or east to China, India, and other Asian countries.[55] For now, the majority of gas goes the Russian route, with the Western route becoming less attractive. Yet the Eastern route is another viable and profitable option, with Chinese investment in pipeline routes and exploration increasing.[56] However, the coercion that may be expected from Russia is not happening in the case of this eastern option. Why not?

Salience of Central Asia

Looking back at Table 6.1, the only state that scores in the moderate category is Turkmenistan. This is because of the lack of Russian allegiance in the form of membership in the Russian-led organizations that would otherwise reduce the salience score, as well as Turkmenistan's role as a transit state and its hydrocarbon endowments. Russia had the monopoly on the transport of Turkmen gas to its European market, but now that has been broken by the Turkmenistan-China pipeline, and it could be further severed if the Trans-Caspian pipeline to Azerbaijan is ever built.[57] This leaves the salience for Turkmenistan and its non-Russian courtships moderately high for the Russian state.

Covered in the introduction, Uzbekistan has only recently embraced Russian hegemonic leadership and joined its sphere of influence. The Western condemnation of President Islam Karimov's actions in the Andijon incident in 2005 has decided this shift in allegiance. President Putin, as well as China's Hu Jintao, was one of the only world leaders to praise Karimov's actions.[58] Since the incident, US troops have been expelled from Uzbekistan, relations with the superpower have soured, and Uzbekistan is joining the Russian-led security and economic organizations. It is currently a member of the Eurasian Economic Community, has been granted probationary membership in the CSTO, and is talking about membership in the Shanghai Cooperation Organization.[59] Therefore, Uzbekistan remains at a salience score of three, with the score being reduced to one once full membership is attained in the near future.

Kazakhstan records a salience score of three for Russia, as not only does it have pipelines and gas, but nearly 40 percent of its population are ethnic Russians.[60] This vast country is an important natural gas link between the other Central Asian states and Russian pipelines, as well as an important supplement to Gazprom's growing supply demands to the European market. Furthermore, Kazakhstan shares a long border with China and has been courted by Chinese national energy companies for it to become one of China's major suppliers of oil and natural gas to feed the growing energy demands for the rising power.[61] However, Kazakh President Nazarbayev, in power since the Soviet breakup, has known the importance to his regime of Russian friendship, and his country is a member of all of the Russian-led organizations in Table 6.1. Kazakhstan is thus rewarded with near-market-price purchases of its natural gas headed for Europe. Its allegiance to Russian institutions may give Kazakhstan the special privileges it has in its shift towards Chinese investment in its energy sector.[62]

The final two Central Asian states in Table 6.1, Kyrgyzstan and Tajikistan, have the lowest Central Asian salience scores. Neither country has natural gas reserves, nor do they serve as important pipeline transit states for Russian natural gas. Furthermore, both countries belong to all three Russian-led organizations. This makes these countries irrelevant for the analysis of this section. What this section has shown is that the countries of Central Asia are much less salient overall relative to those of the Caucasus. The relationship with these countries is somewhat different, as it is Russia that buys the majority of the three Central Asian states' natural gas and pumps it through its pipelines. Therefore, what we expect to see is a relatively high price in the purchase of natural gas from Russia. More on this will be covered in the dependent variables section.

After the attacks on 11 September 2001, the US became dependent on Central Asian states to host its military and play a support role for missions against the Taliban and other terrorist groups. The Pentagon has an air force base near Bishkek, the capital of Kyrgyzstan, and prior to the 2005 Andijon incidents, it had a base in Uzbekistan. Prior to the invasion of Afghanistan, Russia consented to the US establishing a military presence in Central Asia, but only on a temporary basis.[63] Tensions rose as the Shanghai Cooperation Organization called on the US to set a timetable for a complete withdrawal from the region. American drawdown now seems imminent, thus the military presence in the region is near an end. The rivalry for influence in Central Asia between the US and Russia seems to be waning, replaced by a new yet non-rivalrous competition between China and Russia.

This is not to say that the United States and its economic and political reach is completely absent in Central Asia, merely that the salient issues that caused tensions between the United States and Russia have been tempered with the exit of the US military. After the US invasion of Iraq, Putin's acceptance of the US military presence so close to Russian borders as well as his partnership with America in the War on Terror was called into question.[64] Furthermore, Central Asian oil and natural gas reserves were potentially seen as viable alternatives to the ever more volatile Middle East.[65] Therefore, economic, primarily in energy, interest by the US and American energy companies got the attention of Moscow and raised the tensions between the US and Russia over this part of the world. However, as American political influence waned and the authoritarian governments of Central Asia started to assimilate more into Russia's corner, along with the sharp spike in energy prices and the economic downturn of the Great Recession, US energy investment interest has been reduced significantly.[66] It seems that Washington is leaving Central Asia to Russia, and with this enters increased Chinese interest in the energy reserves of this region of post-Soviet space. How has Moscow responded to these increased Chinese energy investment projects in Central Asia? Table 6.10 summarizes the main Chinese agreements with the three Central Asian states that have natural gas: Kazakhstan, Turkmenistan, and Uzbekistan.

The Chinese National Petroleum Corporation (CNPC), the state-owned energy conglomerate, has only recently entered the Central Asian energy sphere. Along with the construction of the Central Asia-China natural

Table 6.10 Chinese natural gas agreements with Central Asian states

	Agreement Name (Year)	Explanation
Kazakhstan	CNPC and Kazmunaigaz sign an agreement to build a pipeline from western Kazakhstan to the east and on to China.	Diverts gas supplies from the Tengiz gas field, on the Russian-Kazakh border, to China over the entire territory of Kazakhstan
Turkmenistan	CNPC signs sales and purchase agreement with Turkmen government (2009)	China will be buying more than 40 bcm per year from Turkmenistan, more than Russia
Uzbekistan	CNPC and Uzbekneftegaz sign an export deal (2010)	China will be importing 10 bcm per year from Uzbekistan, diverting supplies to Russia for China

Source of data: Henderson et. al. (2013).

gas pipeline, which will be discussed in the dependent variables section of this Central Asian analysis, CNPC has agreed to buy natural gas from all three Central Asian states, supplies that take away from Gazprom's import monopolization of natural gas in the region, as well as possible future supplies for the growing European market. It has agreed to invest in the construction of a pipeline that will cross the entire landmass of the vast Kazakh countryside, from the Tengiz gas field in the west to the Chinese border in the east of the country. This gas field is on the border with Russia, and until this pipeline's completion all of the exported gas from this source went through Russian pipelines.

Table 6.10 also shows how China has agreed to purchase significant amounts of natural gas per year from Turkmenistan and Uzbekistan. As the Central Asia-China pipeline has been completed, China now has a reliable supply of natural gas coming from the West. However, this gas has also taken away from the purchasing power of Gazprom, as this gas that could have gone to European markets is now fulfilling China's energy demands. However, there has been little Russian backlash to these Chinese inroads. Cooperating partners in the UN and Shanghai Cooperation Organization, Russia and China have not come to discordant relations over Beijing's increasing presence in Central Asia. According to Kambayashi of *The Economist*, "Russia and China have much riding on their bilateral relationship. The government in Moscow is eager to benefit from its eastern neighbor's economic might, while in Beijing policymakers view Russia as a critical ally on the world stage. All this suggests the two giants will aim to cooperate as much as compete, at least for the moment."[67]

China has been searching to diversify and increase its energy imports in both oil and natural gas. It has been looking to the landlocked countries of Central Asia to feed this growth, and pipelines going east are now either completed and operational or under construction and near completion at the time of this writing. We have found evidence for Russia's potential to be coercive due to recent Chinese investment inroads it has made in this region of post-Soviet space; however, Russia has yet to implement this type of coercive policy. This is because China and Russia are international allies that agree on many geopolitical issues, especially when they involve the United States.

The United States and Russia, therefore, do not have rivalry issues in Central Asia as they do in the Caucasus region. American political, economic, and military influence in the region is becoming scarcer as time goes on. With a war-weary public, American security involvement in the region is likely to be nonexistent, and with the high cost of transporting natural gas and oil out of the region to American or European markets, the prospects of increased American energy company investment is also quite dim. With Russia's rival out of the picture more and more in Central Asia, these states have warmed up to influence from Moscow; however, China has begun to make political and economic inroads with the three states under examination in this section. Therefore, we find evidence here for the assertions made in the second hypothesis of this chapter. Because its relationship with Beijing is crucial, Russia will not coerce states that have made agreements with China, but it will coerce states that work with the United States. The public opinion polls in the next section uncover why Russia may be more permissive over a Chinese presence in Central Asia rather than an American one.

Public Opinion of the Central Asian States and China

This section covers the impact factor of public opinion on Russian coercive energy policy in Central Asia. As the salience scores for Russia and the states of this region are low, and the United States presence in the region significantly reduced, we expect to see public opinion polls with favorable views of the Central Asian states as well as China, an ally and international partner of Russia's. Russian energy coercion in the form of natural gas prices, pipeline transit fees, and pipeline competition, therefore, will be little to nonexistent.

Table 6.11 shows Kazakhstan as Russia's most reliable international partner, according to Russian public opinion when respondents were asked to name three CIS countries. Similarly, Table 6.12 shows that Russians believe the Central Asian state is the most politically stable and economically successful of all the CIS states. Russians see the authoritarian, corrupt, and very opaque government of Kazakhstan as the most reliable, successful, and stable government in the entire post-Soviet space.[68] Traits such as democratic institutions, favorable human rights records, and transparency are not important to the Russian public when it comes to choosing friends of Russia.[69] What is important is close political, economic, and security ties to Moscow, as Kazakhstan is a member of the CSTO, SCO, and Eurasian Economic Community, as well as a free-trade zone customs union with Russia and Belarus. Kazakhstan, therefore, is rewarded with heavy Russian investment in its energy sector and it pays near market prices for its natural gas. However, Kazakhstan has decided to look east and diversify its energy customer base in China. With another major power making inroads into post-Soviet space, it may be expected that the salience for Kazakhstan and the other Central Asian states would rise. Russian outlooks on the rising power via public opinion, however, show that China is viewed as an international

partner and friend to Moscow, unlike the other major power with a presence in post-Soviet space, the United States.
Table 6.13 looks at Russian public opinion about China. During Soviet times, the USSR and China were in a heated rivalry, with different ideological doctrines about communism and its spread around the world.[70]

Table 6.11 Russian opinion of reliable international partners in Central Asia

Respondents asked: Which CIS country (name three) would you call the most reliable partner of Russia in the international arena?

	2009	2010
Kazakhstan	31	37%
Uzbekistan	3	3
Kyrgyzstan	3	2
Turkmenistan	3	2
Tajikistan	2	1

Source of data: http://www.wciom.com/index.php?id=61&uid=154

Table 6.12 Russian opinion of Central Asian states that are the most politically stable and economically successful

Respondents asked: Which of the following countries (name three) do you think are most stable and successful?

	2009	2010
Kazakhstan	29%	34
Turkmenistan	6	5
Uzbekistan	4	3
Kyrgyzstan	5	2
Tajikistan	2	1

Source of data: http://www.wciom.com/index.php?id=61&uid=358

Table 6.13 Russian opinion of Russian-Chinese relations

Respondents asked: How would you assess the current China-Russia relations?

	2005	2007	2009
Friendly	15%	19	17
Good, good-neighborly	19	17	17
Normal, quiet	40	40	39
Cool	11	10	10
Tense	3	3	5
Hostile	2	1	1
Hard to tell	10	10	11

Source of data: http://www.pewglobal.org/database/indicator/24/country/181/

Today Russia and China see each other as the counterweight to American dominant power throughout the globe, and working together as allies is seen as beneficial to both Moscow and Beijing. This post-Cold War friendship is translated in the results of public opinion polls (in Table 6.13) that ask about current Russian-Chinese relations. Roughly 35 percent of Russians see them as friendly or good with another 40 percent seeing them as normal or quiet. Only around 5 percent of Russians see the current relationship with China as tense or hostile. These results are in stark contrast to Russian public opinion about the United States, presented in the previous public opinion section of this chapter, where opinions of animosity are high. Therefore, a Chinese presence in post-Soviet space will not lead to Russian coercive energy policy with states that engage with China over natural gas export agreements.

Table 6.14 shows the opinions of Russians when asked about the future of Russian-Chinese relations in the 21st century. As the results show, nearly half of Russians see China as either an ally or close partner to Russia in the foreseeable future. Roughly 30 percent see the future of relations being competitive or rivalrous, with a sizeable 21–24 percent believing it to be hard to tell at this point. What is clear is that there are much more optimistic appraisals about relations with China than there are with the United States. Therefore, as the theory of situational coercive diplomacy predicts, other states that are not rivals of Russia gaining influence in post-Soviet space over natural gas issues will not lead to an increase in coercive energy tactics if Central Asian states engage with China.

This section has presented how Russian public opinion polls over issues of Central Asia and the major power, non-rival China, are in stark contrast to the public opinion polls of the Caucasus states and the United States. Along with the low salience scores of most of the Central Asian states for Russia, there should be favorable natural gas

Table 6.14 Russian opinion about the future of Russian-Chinese relations

Respondents asked: In your opinion, will China be Russia's friend or enemy in the 21st century?

	2005	2007	2009
Ally, friendly nation	22%	28	20
Close partner	26	24	27
Dangerous neighbor, competitor	25	20	24
Rival, enemy	6	4	5
Hard to tell	21	23	24

Source of data: http://www.pewglobal.org/database/indicator/64/country/181/

policies by Russia with these countries. Before looking at this evidence in natural gas and pipeline pricing and pipeline competitiveness, first we must look at the fourth impact factor—Russia's great power identity in Central Asia. Relations with Central Asia are less important in Russia's image as a great power to itself and the international community than the European and Caucasus regions of post-Soviet space, keeping issues of energy in this region at low salience.

Russian Great Power Identity in Central Asia

Unlike in the Caucasus region, Russia's historical ties to the Central Asian countries are not as long nor do they have important core implications for the development of the Russian Empire, Soviet Union, and modern Russian state.[71] Primarily Muslim and secular, the region does not fill the ethno-religious, Slavic-Orthodox identity issues that are very salient to the Russian elite and masses, as the European and Caucasus regions do. Russian influence in the region began in the late 18th century when the Kazakhs asked for Russian support from the invading Uzbek khanate tribes.[72] Russia answered the call and eventually took over and colonized the entire region by the mid-1800s. It was conquered militarily between 1865 and 1884. The region gave the Red Army some resistance during the Russian civil war years of the early 1920s, and in 1924 the five entities that are today the five states of Central Asia were created.[73] With many revolts throughout the 1920s, the peoples of this region finally capitulated to Stalinist Russian rule in the early 1930s. These five entities became states for the first time in 1991.

The conception of a Central Asian region of similar peoples and political interests is, according to Nygren, "to some extent 'artificial,' and the very notion of Central Asia is difficult to place even geographically; it might be soon as part of Asia, part of the (greater) Middle East—its most northern tier, and it could have (but did not) become an arena for a great game between Iran, Turkey, Russia, and China after the demise of the USSR."[74] Immediately after the Soviet collapse, Russia did not see Central Asia as an important region politically, economically, or militarily. The Yeltsin Administration concentrated on privatization, democratization, and integration with the West, and Central Asia had no place in the fragile Russian state. Central Asia "was basically lost to Russia, mostly because of its lack of capacity to deal with the new states. Russia's economic relations with Central Asia stagnated, trade fell sharply and investments were close to zero."[75]

It was not until the terrorist attacks on 11 September 2001 that Russia's interests in Central Asia rose sharply. Realizing that the borders of

Central Asia were porous and not adequately defended against terrorist and insurgent groups, Russia allowed a significant American military force to set up bases in the Central Asian states of Uzbekistan and Kyrgyzstan and also allowed American aircraft to fly over Russian airspace to conduct its military operations against Al Qaeda and the Taliban in Afghanistan. This very rare Russian-American cooperation did not last long and came to an end with the US invasion of Iraq in 2003, which Russia vehemently opposed, and the American reaction to the events in Andijon in 2005.[76] Uzbekistan expelled the Americans from bases in their country shortly thereafter, and Russia voiced its displeasure with the continuing US presence in Kyrgyzstan. With the US campaign in Afghanistan drawing down, the US-Russian rivalry in this region is also being relaxed. Russia has close political ties with all Central Asian states except Turkmenistan, the closed state that has no close ties with anybody.[77] With their principal rival out and their key major power ally in China in, Russian salience of the region is now reduced, so coercive energy policy will not take place.

This section has shown that Russia's great power identity is not as tied to the Central Asian region as it is to the Caucasus region. The Central Asian region has only been in Russia's sphere of influence for about 150 years, and its peoples are mainly Muslim in religion and Turkmen in ethnicity.[78] It has been an outpost of the Russian Empire and later Soviet Union and does not hold the same importance for the reassertion of Russian power for the new post-Soviet Russian state. Although the Russian language is used for cross-border communication, its use as the widespread second language vernacular for the population has been greatly reduced since the USSR's fall.[79] It has only gained geopolitical salience for Moscow when the United States showed interest in the region after 9/11. With the Americans leaving, Russia has seen voluntary allegiance to Moscow by most of these states, with Kazakhstan, Tajikistan, and Kyrgyzstan full members of the Russian-led IOs presented in this chapter. Turkmenistan is part of no political organizations of any kind and keeps its relations with other states to a minimum, with its major partners only being Russia, Iran, and China. Therefore, the salience of the region for Russia remains low, the US-Russian rivalry in the region is near extinct, a new major power in China is considered an ally of Russia, and the Russian public opinion of the states of this region is favorable. These factors and the lack of need for Central Asia to prop up a great power identity will lead to the lack of Russian coercive energy policy in the region. The next section shows evidence for this assertion and presents the natural gas pricing and pipeline fees, as well as the lack of competition for pipeline routes in the region.

Central Asian Dependent Variables: Russian Rewards and Pipeline Politics

Table 6.15 shows the natural gas prices Russia pays its Central Asian customers of Kazakhstan, Uzbekistan, and Turkmenistan for gas headed for Gazprom's European customers as well as Ukraine, and it shows the prices China and Iran pay Turkmenistan for natural gas via the new Turkmenistan-China pipeline. In the early years, Russia bought gas from these countries at subsidized prices, usually because there were barter contracts that were very opaque, and energy export developments were at a very low level.[80] Gazprom began to change this pricing scheme with heavy investment in exploration, pipeline repair, and extraction.[81] As demand from the European market grew, Gazprom needed other, more reliable sources of natural gas that could be transported through its pipelines. As the Caucasus was looking west in that it was looking towards Western investment and Russia-circumventing pipeline projects, the landlocked countries east of the Caspian Sea in Central Asia were the perfect supplement to this growing supply need. As compliance and allegiance to the Russian state was required to keep this steady supply going, a slow yet steady increase in prices was instilled throughout the decade, and by 2011 all three Central Asian states were being paid near market prices for their gas.

Any gas sold to Ukraine by these three states must go over Russian territory, and after the January 2006 Russo-Ukrainian gas dispute, no separate bilateral contracts between Ukraine and Central Asian states was permitted; all prices to Ukraine are now set by Gazprom.[82] Therefore, the high prices these states charge Ukraine are due to Russian coercion and not the decisions of the energy elite of the three states. What is clear here is that compliance with Russian interests has led to

Table 6.15 Natural gas pricing in the Central Asian region ($ per mcm)

Central Asian State	Export Scheme	2003	2004	2005	2006	2007	2008	2009	2010	2011
Turkmenistan	To Russia	44	44	44–60	65	100	130–150	340	200	260
	To China	-	-	-	-	-	-	-	180	240–245
	To Iran	42	42	42	65	75	130–150	120–160	120–160	200–220
	To Ukraine	44	44	58	-	-	-	-	-	290–295
Uzbekistan	To Russia	-	-	42	60	100	130–160	190	200	250
	To Ukraine	51	-	64	95	-	-	-	-	290–295
Kazakhstan	To Russia	-	-	47	60	100	180	180	170–190	250
	To Ukraine	-	-	-	-	-	-	-	-	310–315

Source of data: Henderson et al. (2013).

a huge increase in revenue for these poor states of Central Asia, and the continuation of selling gas to Gazprom is now in these states' best interest. However, Turkmenistan, a state with no political allegiance to any major powers and with a policy of permanent neutrality, has begun diversifying its supply, as it has had Iran as a customer to serve the northern provinces of this Islamic state, and now it has a powerful, wealthy, and energy-starved customer in China. Deliveries of gas through the Turkmenistan-China pipeline, also known as the Central Asia-China pipeline, began in 2009. The pipeline is now being expanded to include deliveries from Kazakhstan and Uzbekistan, and now Russia has a new competitor in Central Asia. However, due to the impact factors for Central Asia presented in the previous sections, Russia's non-coercive and generous natural gas buying processes for these countries have continued.

Table 6.16 shows the existing natural gas pipeline systems in Central Asia. As is evidenced, the competition over routes to the European markets is nonexistent. Russia still holds the monopoly on transit routes to Europe from the region, and it seems that this will be the case for a long time to come. The Central Asia Center system is an updated version of the old Soviet routes, and most of the exported gas from the region travels to Russia and external markets. The Bukhara-Tashkent-Bishkek-Almaty pipeline is a regional one that supplies gas from Russia, Uzbekistan, and Kazakhstan to the have-nots of the region, Tajikistan and Kyrgyzstan.[83] The pipeline from Turkmenistan to Iran is a smaller operation that was set up so that the northern provinces of Iran could import cheaper gas, as the costs of transport from Iran's own gas fields in the south of the country were too high. The Turkmenistan-China pipeline is the newest pipeline that became operational in 2009 and has broken the monopoly of

Table 6.16 Existing Central Asian natural gas pipelines

Pipeline	States on Route
Central Asia Center (CAC)	Turkmenistan-Uzbekistan-Kazakhstan to Russia
Bukhara–Tashkent–Bishkek–Almaty	Uzbekistan-Kazakhstan to Kyrgyzstan-Tajikistan
Turkmenistan-Iran	Turkmenistan to Northern Iran
Turkmenistan-China (Central Asia-China)	Turkmenistan-Uzbekistan-Kazakhstan to China

Source of data: Dellecker and Gomart (2011).

Gazprom's ownership of Central Asian pipelines. Uzbek and Kazakh supplies are soon to join their Turkmen counterpart and begin supplying China as well.[84] However, Russia is getting what it wants in a reliable supplementary supply to Europe, and as long as this supply does not dwindle, the status quo of near-market purchases from Gazprom will continue.

Table 6.17 lists the pertinent proposed natural gas pipeline projects that would travel through Central Asian territory.[85] The Caspian Coastal pipeline will transport Turkmen gas across the least amount of territory possible—from the eastern gas fields to the Caspian coast and up the coastline through Kazakhstan to Russian pipelines headed towards Europe.[86] This will only increase supplies for Russia and Gazprom and keep Central Asia loyal to Russia's monopoly over transit to Europe. The Tajikistan-China proposal has been endorsed by Putin and will be paid for largely by China's national petroleum company; it will transport Turkmen gas through Tajikistan on to China as another outlet to feed China's growing gas demands. Finally, the Trans-Caspian pipeline, discussed in the previous section, looks close to dead due to a lack of Western investment as well as Turkmenistan's eastward-looking projects. Therefore, the competition over pipeline routes in Central Asia, existing or proposed, is very little.

This section's evidence has shown the contrasting energy policies of Russia and Gazprom in Central Asia when compared to that in the Caucasus region. The low salience of the region; the relative absence of Russia's continued principal rival, the United States, in the region; the non-adversarial public opinion of Russian citizens about these countries as well as China; and the lack of attributes of the region for Russia's great power identity have all led to less and less energy coercion, as all countries that sell natural gas to Gazprom are now receiving near market prices. Furthermore, the competition over pipeline routes to Europe in the region is nonexistent, with the only competing pipeline routes headed for China, Russia's perceived friend and ally. Russian energy coercion in Central Asia, therefore, is at a very low level.

Table 6.17 Proposed Central Asian natural gas pipelines

Pipeline	States on Route
Caspian Coastal	Turkmenistan-Kazakhstan to **Russia**
Tajikistan-China	Turkmenistan-Tajikistan to China
Trans-Caspian	Turkmenistan to Azerbaijan

Assessment and Conclusions

This chapter has presented empirical evidence to back the theory of coercive policy by comparing two regions of post-Soviet space: the Caucasus and Central Asia. The impact factors of issue salience, rivalry, public opinion, and identity have predicted whether or not Russian coercive energy policy will be used in the form of gas pricing and pipeline fees as well as competition over pipeline routes to external markets.

In the Caucasus, the salience of this region is high, the presence of Russia's principal rival (also covered in the next chapter), the United States, in the region is also high, the public opinion of the states of the region and the United States is adversarial, and Russia's identity as a great power requires that the region is well within its sphere of influence, with no outside powers making inroads. Russia has even resorted to armed conflict with Georgia in an attempt to lessen the Western turn the country had been making and to discredit its reputation as a reliable transporter of natural gas and oil that circumvents Russian territory. Therefore, we see fluctuating prices for natural gas and pipeline fees as well as high competition over pipeline routes between Russia and Western companies. Russian coercive policy is present in the Caucasus region.

For Central Asia, Russia sees the region as less salient, the United States' presence is limited, public opinion about the region as well as China are not adversarial, and Russia's great power identity is not tied to the history and peoples of this region. Russia has found more reliable partners in this region with this dynamic; therefore, it imports the natural gas from the three countries of Turkmenistan, Uzbekistan, and Kazakhstan at near market prices to supplement the growing European demand. Furthermore, these states have diversified their exports with China, and Russian coercion has not come about with the entry of this outside power into post-Soviet space. Therefore, there is ample evidence for the theory of situational coercive diplomacy with energy for the findings of this chapter.

We therefore fail to falsify all of the hypotheses presented in this chapter. The first hypothesis states that when energy salience is high for a region, as it is in the Caucasus, Russia will be more likely to use coercive energy tactics; while if the energy salience is low, as it is in Central Asia, more accommodating energy policy is likely to be present. The second hypothesis states that when the United States is present in a post-Soviet region in the form of energy investment, Russia is more likely to use coercive energy policy; this hypothesis also fails to be falsified. Again, when comparing the two regions, Azerbaijan sees heavy American investment while China has been making investment inroads in the Central Asian states of Turkmenistan, Uzbekistan, and Kazakhstan. Russia has been coercive in the Caucasus but not in Central Asia. The presence of rivalry in post-Soviet space, therefore, plays a factor in Russia's coercive energy tactics.

Public opinion polls show that Russians approve of coercive tactics in post-Soviet space, which may justify some of the action that Russia takes with its energy power in the region. Furthermore, the Russian public sees the United States and Georgia, two states heavily involved in anti-Russian energy policies in the Caucasus, as Russia's two most hated enemies. On the other hand, most Russians see China and Kazakhstan, two states with energy interests in Central Asia, as Russia's closest friends. Furthermore, Russia sees the Caucasus region as essential to the great power identity for 21st century Russia, with long historical, ethnic, and cultural ties to the region. Central Asia, on the other hand, has not been historically tied to Russia as long and does not have the same ethnic and cultural ties to Russia. Therefore, the third and fourth hypotheses of this research project also fail to be falsified. Next we look at a situation where Russia is accommodative and cooperative with other great powers in another region of growing importance for Russia—the Arctic.

7
Russian Foreign Policy in the Arctic Region: Issues, Preferences, and Conflict Moderation

Introduction

In August 2007, a Russian expedition sent an unmanned submarine to the bottom of the Arctic Ocean and planted a Russian flag on the geographic North Pole. Canadian Foreign Minister Peter Mackay gave his opinion shortly thereafter: "This isn't the fifteenth century. You can't go around the world and just plant flags and say 'we're claiming this territory.'"[1] Echoes of this sentiment were heard from Western politicians in the United States and Europe. Russia is now deeply invested in the Arctic region, and the question we ask in this chapter is what sort of dynamics will this evolving situation lead to? What does it portend for Russian foreign policy?

Brash reactions by the West over Russia's symbolic achievement imply that territorial claims in the Arctic Ocean will begin to manifest, which could in turn lead to a literal cold war in the Far North between Russia and the West. The rapidly melting sea ice due to global warming is changing geopolitical and economic realities of the northernmost region of the globe. While every state in the world has promised to not claim territory in Antarctica under an international treaty, it is unclear what relationship states will develop in the Arctic, especially in the context of developing sea lanes and the possibility of massive amounts of natural resources in the region.[2] Will states claim territory both above and below the melting ice in the Arctic, and will these developments change or alter the course of international affairs?

Finger pointing over flag planting by Western media and leaders suggests that Russia is up to its normal foreign policy behavior, using power politics and displays of power in an attempt to get what it wants in the name of the national interest.[3] The Western response to this Russian achievement also seems to be typical with quick dismissals and accusatory remarks over Russia's ambitions. Russian Foreign Minister Sergey Lavrov explained the flag planting from his country's point of view: "Whenever explorers reach some sort of point that no one else has explored, they plant a flag. That's how it was on the moon, by the way."[4]

Russia has reached a new apex of its power in the post-Cold War era. So far in this book, we have discussed how Russia has used its new forms of power in the territory of the former Soviet Union, the question is if this form of situational coercive diplomacy will be extended to geopolitical interactions in the extreme north. The goal of this chapter is to examine Russian geopolitical power and foreign policy in the seas through a case study of the Arctic. We posit that outside of post-Soviet space, the possibilities for peace and cooperation are more likely than any path to conflict. The maritime operations in the Arctic are different from those in the former Soviet Union because the actors in this region are not traditional Soviet vassal states, and the potential territorial disagreements are not symbolic and transcendent but concrete and divisible. What is striking about this arena is that the Russians are willing to use international institutions and legal processes to assert foreign policy perspectives. The status of Russia is not in question in the Arctic, what is open to debate is the global reach and power projection capabilities of a reemerging Russia. Here, we find that Russia is not using its new forms of power but rather maneuvers to utilize international institutions and legal statutes.

The Stakes under Contention: Arctic Territoriality

As covered in Chapter 2, disputes over territory are very salient to states because of their potentially symbolic properties. There are disputes over territorial claims between Russia and the West in the Arctic, as well as among the Western countries themselves, and because these disputes are territorial in nature, they could be dangerous to global security. The issue-based approach has succeeded in theoretically and empirically demonstrating that most conflicts originate with a territorial dispute.[5] Once territorial disputes arise, alliances can form, arms races can proliferate, rivalries can emerge, and domestic hardliners may come to power, all of which increase the probability of armed conflict.[6] However, it is the type of territorial dispute and how territorial disputes are handled that can lead a pair of states down the path to either war or peace. In other words, there are steps out of war for territorial disputes.

As highlighted throughout this book, Russia uses coercive power politics tactics to express foreign policy preferences in the former Soviet Union, but here we identify mechanisms that are avenues towards peace for the issues under contention in the Arctic. The stakes are relatively low for the disputed claims within the Arctic. The three main issues under contention are the extraction of natural resources (primarily hydrocarbons), shipping lanes, and fishing waters that recently became available due to advances in technology and global climate change. Some contend that climate change will lead to conflict, but we demonstrate here that the potential for conflict really depends on how contentious issues are handled.[7] The location dynamics alone, with the Arctic being so far from the homeland of each state involved, make it likely that conflict will be muted in the region. In the end, this outcome still depends on how states choose to behave.

The "Arctic Five," which is the popular term for the five states that have coastlines bordering the Arctic Ocean—Russia, the United States, Canada, Norway, and Denmark via Greenland—are leading the way in the race to grab resources and opportunities in the region.[8] Thus far, we have shown that depending on the context of the situation, Russia prefers coercion to achieve its ends over issues it finds salient. However, due to the nature of the stakes for the Arctic territorial issues, we argue in this chapter that there is an avenue towards peace and cooperation with Russia and the other four Arctic states. Despite the presence of territorial disputes and issues that seem to be at the core of the Russian national interest, there is nuance to be found in the territorial dispute literature. A state must have a transcendent territorial dispute for it to be of an intractable quality. As we argue in Chapter 2, the dispute must also be connected to public expressions of what the national interest might be, and Chapter 4 demonstrates that the Arctic is not considered a core security issue for the Russians. Finally, there is a rivalry between Russia and the United States, as we document in Chapter 3, but this rivalry mainly operates in the former USSR over regional issues and has not yet extended to the Arctic. The nature of the issues under contention in the Arctic demonstrates that Russia can be dealt with more diplomatically and with less brinkmanship, and international security concerns, at least with Russia, can be eased.

The Disputed Region

The effects of climate change are being felt particularly acutely in the polar regions. Scientific data suggests that pollutants by industrial and industrializing states are trapping greenhouse gasses in the atmosphere, effectively disrupting the natural heating and cooling processes of the oceans.[9] Due to the amount of ice and snow that reflects sunlight, the Arctic Ocean is especially feeling these effects. The icepack in the extreme North is melting at an alarming rate. As the ice melts annually, less sunlight is reflected back to space by this usually huge area of white surface, which spirals into a positive feedback loop that melts the ice more and more each year.[10] This warming is also melting the permafrost in the landmasses north of the Arctic Circle. Local wildlife are losing their traditional feeding and hunting grounds.[11] Indigenous peoples are losing their traditional ways of life. Some extreme projections contend that the Arctic Ocean will be ice-free by 2050.[12]

Besides the potential of rising ocean levels and disrupting the natural cycle of seasons in the future, the icepack melt is also allowing for economic and territorial access, which has been long impossible due to cold temperatures and ice. Economic access means that ships without icebreaking technology can now use the region for longer periods, thus shipping lanes can be used where they could not before, offshore drilling is possible because of the reduction in destructive icebergs in the region, and seaports in the Far North are now more useable for longer periods. This new economic potential leads to disputes over shifting land-based territorial claims and continental shelf claims, new shipping lanes, and whether underwater exploration should be made legal.

Melting ice in the Arctic Ocean will allow ships without icebreaking technology to roam where only specialized and expensive ships could roam before. Cargo ships will have shortened routes to their destinations at the expense of the Panama and Suez Canal routes. The fabled Northwest Passage and Northern Sea Route are quickly becoming a reality. Fishing ships will have new waters to bring in more varied catches of fish and shellfish. Platforms to discover oil and gas will be able to be built in locations where the shifting ice made this type of construction impossible only years ago.[13] These new economic realities could lead to competition for sovereignty, border disputes, and claims on sea lanes. These points of contention are growing between the Arctic Five as well as other states and organizations wishing to get in on the opportunities. China, Japan, and the European Union have also expressed interest in the Arctic's fate.[14]

Scientists project that the Arctic seabed contains vast reserves of untapped oil and natural gas. Early returns from the US Geological Survey (USGS) suggest that the Arctic could hold the last large remaining oil and natural gas reserves on the planet.[15] Optimistic projections indicate that the Arctic could hold up to 25 percent of the world's undiscovered oil and gas deposits.[16] Furthermore, the Arctic's long continental shelf could lead to the potential for commercially accessible offshore oil and gas resources, with the largest deposits off the coast of Russia.[17] The territory Russia is claiming, as well as the newly accessible land-based reserves due to massive thaw of tundra lands, could also contain 25 percent of the world's hydrocarbon reserves (approximately 15.5 billion tons of oil and 84.5 trillion cubic meters of gas).[18] As the reserves of these hydrocarbons in temperate zones such as the Middle East become scarcer or politically unavailable, competition for extraction rights in the Arctic Ocean has the potential to bring conflict to the polar region. Territorial disputes could become more acute, leading to power politics practices by states, which could raise the possibility of military settlements rather than peaceful ones. Yet it is important to look at how territorial disputes are handled before we declare that war is imminent in the Arctic Ocean.

Just because a territorial resource dispute can become contentious does not mean that this outcome will occur. Huth demonstrated early on that resource-based claims were the least likely type of territorial claim to lead to war.[19] This trend has continued despite doomsday prognosticators who warn of coming resource conflicts.[20] We argue here that the nature of disputes in the Arctic region, although territorial, can be handled peacefully through the mechanisms of international institutions, regimes, and law because these disputes are not symbolic in nature, a situation that typically leads to outright conflict. The hype of a Russo-Western conflict projected by pundits and policymakers is unwarranted.[21] Through the lens of the issue-based approach, we argue that the Arctic issues can be resolved through bargains where everyone gains, as the salience of the Arctic issue is not high for Russia and the four Western Arctic states.[22] The Arctic issue will not be tangled up in rivalrous, intangible, and escalatory disputes characteristic of other regions.

We show through an in-depth issue-centric case study that Russia is using international regimes and institutions to claim what it perceives to be its rightful territory. It continues to work with institutional bodies and regimes even though it ignores these actors in other regions or creates its own parallel structures. The structure of this case study is based on three critical areas outlined in Chapter 2: 1) salience of the disputes to each of the five countries, 2) the particular location of the Arctic region as well as the timing of this issue on the global agenda, and 3) the public opinion of the populace of the Arctic Five states. Using these questions to focus our analysis, we find that conventional wisdom is often wrong.

Russia and other states are unlikely to escalate Arctic disputes in a meaningful manner anytime soon.

Institutions in the Arctic

The current Arctic ice melt is causing a lust to acquire potential natural resources, such as oil and natural gas, new shipping lanes for cheaper commerce, and new fishing grounds in the previously non-navigable Arctic Ocean. Every Arctic Five state is aware of these possibilities. Russia, it seems, is taking the lead in mobilizing for the future of a nearly ice-free Arctic. Besides the supposed power play and symbolic grandstanding of the flag planting on the North Pole, Russia has made territorial claims to the Arctic seabed based on the UN Convention on the Law of the Sea (UNCLOS).[23] Except for the United States, the Arctic Five states are signatories to this convention. The US Senate has blocked ratification of this treaty numerous times, even with willing presidents, George W. Bush and Barack Obama, as well as the entire military urging its approval. A small group of senators with the power to filibuster, or block legislation, believe that signing on to this convention would be an infringement upon American sovereignty, noting "America's special role in the world."[24] The convention indicates that a state's territorial waters extend 12 miles off the coast, and the Exclusive Economic Zone (EEZ), where a state has exclusive rights to extract resources from its continental seabed, is set at 200 nautical miles. Another stipulation of this convention is that a state has the right to argue for an extension of its EEZ if the continental shelf extends beyond 200 nautical miles, up to and including an additional 150 miles.[25] In other words, a state has the right to the exclusive extraction of resources of seabeds up to 350 miles off of its coast, if it makes a good argument. Russia has submitted this maximum territorial claim as well as claims that are well beyond the reach of the rules of UNCLOS. Russia argues that the Lomonosov Ridge is an extension of the Eurasian (Russian) continent and it thus has legitimate claims to much of the Arctic. These claims are beyond the practices of the traditional international norms, which could be seen as a power play by the other Arctic Four states. These unorthodox claims by Russia suggest that the Arctic issue area is leading to a situation where peaceful settlement and compromise will not be possible. The spirit of the Arctic Council, however, suggests otherwise.

Originating in 1996, the Arctic Council is comprised of eight states and various indigenous Arctic peoples with interests in the fate of the Arctic. These states include the Arctic Five as well as Sweden, Finland, and Iceland.[26] There are other "observer" states in this council, including China, Japan, and many EU countries.[27] The purpose of the Arctic Council is for a "high-level intergovernmental forum to provide a means for promoting cooperation, coordination and interaction among the Arctic States, with the involvement of the Arctic Indigenous communities and other Arctic inhabitants on common Arctic issues; in particular, issues of sustainable development and environmental protection in the Arctic."[28] The five states with coastal borders, however, are the ones that have been participating in most of the discourse since the council's creation. Although the Arctic Council is mainly a forum for discussion, legislation and written agreements have occasionally come out of the body. An example is the agreement by all five governments that the UNCLOS rules apply to the Arctic, which is helping mitigate and reduce the stakes of the Arctic from becoming races for resources.[29] Even the United States, although not a signatory to the UNCLOS, has agreed that it will adhere to the rules set forth by the convention.

Another agreed-upon stipulation of the Arctic Council is the adherence to the decisions of a body within the UNCLOS, called the United Nations Commission on the Limits of the Continental Shelf (CLCS). Not directly affiliated with the Arctic Council, this commission is a review body of scientists from around the globe created under the UN convention.[30] The Arctic Five have allowed for their territorial fates to be given to this small group of academics. Russia was the first to submit its territorial claim to the underwater shelf in the Arctic to this commission. Russia's claim is beyond the 350-mile limit that the convention allows; however, Russia believes that since the Arctic continental shelf extends out an unusual distance, this unusual claim is warranted. Therefore, Russia is claiming nearly half of the Arctic underwater territory, including the geographic North Pole.[31] Canada, Denmark, and Norway have announced their intentions to submit their territorial claims to CLCS. Using the data submitted by these countries as well as their own research and data collection, the CLCS will decide if the territorial claims by countries are legitimate. Russia has already announced that it will adhere to the decision made by this commission.[32] The only detractor to following international law for territorial claims is the United States. It has not signed on to the UNCLOS, thus has no right to submit claims to the CLCS. This commission of scientists has the authority and the respect of states as the final authority on territorial claims in the Arctic. Therefore, this commission is an important mechanism in settling territorial claims and mitigating disagreements in a peaceful manner.

Perhaps another important institution in the Arctic is the one started by the European Union and is commonly known as the "Northern

Dimension." This organization is comprised of the EU, Iceland, Norway, and Russia. The organization supports the same spirit of cooperation that the Arctic Council does, without the territorial verbiage.[33] We see this organization as the EU's attempt to be relevant in the future of the economic potential of a melting Arctic Ocean. As we discover in the following pages, we find that the EU and most states of the Arctic Council do not have the same outlook on the fate of the Arctic. The Arctic Council, which is the organization that is comprised of the states whose territories actually border the Arctic Ocean, is the more relevant organization to examine for the purposes of this chapter. As discussed later in the chapter and throughout the book, territorial issues tend to be the most salient, and examining the organization that is capable of setting these salient issues will be the focus here.[34]

Russia's economy is based on energy exports, and the fact that more energy resources are becoming available due to global warming can only enhance the Russian share of the international energy market.[35] Global warming also means that much of Russia's territory currently under permafrost will eventually thaw. This will lead to more arable land for farming, grazing, and human habitation. It will also make drilling for land-based natural resources in these permafrost regions much easier to accomplish. Although potentially disastrous for many low-lying parts of the world, climate change in Russia could reap it huge economic benefits. Russia has much to gain north of the Arctic Circle, while Western states are rushing to meet the challenge and sustain their current levels of productivity.

Western states are taking the threat to their interests in the Arctic seriously.[36] Beginning in the winter of 2009–10, the United States, Canada, and Denmark have been participating in military drills in the Arctic, known as Operation Nanook.[37] Furthermore, the US Defense Department has reached an agreement to train American soldiers and sailors in northern Norway as of August 2010. Huebert points out that "for the past 20 years, none of these four states saw a need to exercise their forces in the Far North. They may be telling their citizens that all is well in the Arctic, but their actions suggest that this is not what they truly believe. . . .It's hard to avoid the conclusion that Moscow is the target of these vigorous military exercises in the Arctic. And if it's not Russia, something certainly is poking the four northern NATO allies in the side."[38] We do not subscribe to this view, as while the potential threat of Russia always looms, as noted in Chapter 5, it is unclear if the states are just being prudent in preparing for new global missions or training to deal with Russia in a conflict. Preparation does not necessarily mean that conflict is a given or even desired.

The idea of another "frozen conflict" between the West and Russia is an exercise in hyperbole. There is hope for cooperation and peace via the Arctic Council and UNCLOS. This body has been relatively successful in getting Russia and its potential adversaries to agree and work together in taking advantage of the new economic opportunities in the Far North. Using the four issue measures of salience, location, timing, and public opinion, we find that conflict in the Arctic is a remote possibility, contrary to popular belief.

The Probability of Issue-Based Conflict in the Arctic

Using the issue-based approach and our framework (Chapter 2), we posit that the Arctic issue area should be measured using four tools of analysis: salience, location, timing, and public opinion.[39] From these areas, we derive three hypotheses to examine in the Arctic region:

> H1: *Territorial claims over natural resources, primarily oil and natural gas, will be of low salience because of international norms, the Arctic's remote location, the timing of the Arctic issue relative to other issues on the international agenda, and the elite and public opinions of the states bordering the Arctic region.*
>
> H2: *Territorial claims over emerging shipping lanes will be of low salience because of international norms, the infrequent use of these lanes, the small number of states using these lanes, the timing of the Arctic issue relative to other issues on the international agenda, and the elite and public opinions of people in the states bordering the Arctic region.*
>
> H3: *Territorial claims over emerging fishing waters will be of low salience because of international norms, the small amount of disputes over fishing territories, the Arctic's remote location, the timing of the Arctic issue relative to other issues on the international agenda, and the elite and public opinions of people in the states bordering the Arctic region.*

An issue area involves contention among actors over proposals for the disposition of stakes among them, as well as the issue dimension, which is the manner in which actors perceive and define the issue before them.[40] In contrast to power-based theories, looking at issue areas and how salient states perceive them to be through the issue-based approach will be able to predict foreign policy outcomes more accurately. Power-based theories assert that because State A is more powerful than State B, State A will get its issue preference over B. As this scenario does not always hold empirically, looking at issue areas and how salient State A finds them in relation to State B could explain who gets what in the international arena better than power-based assertions.

Mansbach and Vasquez give four basic issue areas that have different means and ends: issue areas that have intangible ends and intangible means, issue areas that have intangible ends and tangible means, issue areas that have tangible ends and intangible means, and issue areas that have tangible ends and tangible means.[41] Issue areas with tangible ends means that the stakes can be divided up, that is, they are observable or physical. Territorial issues are usually able to be divided up tangibly. Usually there are international regimes that can lead to tangible means, as states have agreed upon ways of dividing the stakes so that every contender can come out a "winner." Intangible ends mean that the stakes are more abstract; they are hard to make divisible and are not physically observable. An example of this may be the democratization of a region or state. Actors may have different definitions of democratization, and it may be hard for all contenders to be satisfied. Intangible means denotes how the stakes are to be handled or divided. For many territorial disputes, although the ends are tangible in that the territory is divisible and physical, the feelings associated with these issues are usually intangible. States often see the territory as symbolic, part of the homeland, and will be willing to fight for the whole territory because of its symbolic nature.[42]

The claims in the Arctic, although territorial, are either symbolic or heavily populated, and they are divisible. Therefore, the territorial issues in dispute in the Arctic take a path where the tangible ends are pursued with tangible means. When this type of foreign policy is used, conflict is less probable and accommodationist strategies are usually the norm.[43]

We assert that the Arctic and its potential for extraction of resources (mainly oil and natural gas), new fishing waters, and shortened shipping lanes for worldwide markets make the stakes concrete and thus tangible.[44] "For the most part, actors contend for stakes in the belief that access to them will afford immediate value satisfaction. These are concrete stakes."[45] Therefore, when the stakes are divisible in some finite way, they are tangible. Tangible stakes can be resolved by equally dividing up the spoils in some agreed-upon way. Intangible stakes are usually not divisible and thus not concrete. Thus, the salience of the Arctic issue and the tangible stakes involved for each of the Arctic Five states, along with the particular isolated location of the Arctic, the timing of this issue area in regards to others on the current global agenda, and the public opinion of the five states are all contributing factors to the disputes over territory in the Arctic being resolved in a peaceful manner.

Salience

Salience is a key factor left out of the analysis in many examinations of international affairs. "The salience of an issue is the degree of importance attached to that issue by the actors involved."[46] It is just as important to ask how much a good is valued to you as it is to ask what goods you want. Issues perceived to be highly salient to both sides will often lead to armed conflict, while less salient issues will usually end in third-party mitigation, negotiation, and peaceful settlement.[47] It is the highly salient disputes that are troubling for the international system.

The question we answer in this section is how salient is the Arctic issue to the five states vying for territorial claims for the extraction of natural resources, fishing water rights, and sea lane passages? A measurement of issue salience is tricky, yet scholars have attempted and succeeded in giving us guidelines on how to measure the importance of issues to particular states. We again use and alter Hensel's scale for the salience of territorial issues for the issues of the Arctic.[48]

Remembering the previous chapter, the six factors measuring issue salience for territorial claims are 1) the presence of a permanent population, 2) the confirmed or believed existence of valuable resources, 3) the strategic economic or military value of the territory's location, 4) the existence of a state's ethnic and religious kinsmen in the territory, 5) whether or not the territory is considered part of the homeland or is a dependency, and 6) whether or not the territory is part of the mainland or is offshore.[49] We keep these measurements intact and apply them to the Arctic region for each state that has stakes in this part of the world.

As presented in Chapter 6, Hensel and Mitchell incorporate other factors contributing to either peaceful settlements or escalating conflict; they are issue management, past interactions, institutional context, and the characteristics of the "adversaries." We add these factors into our analysis to get a more complete picture of issue salience. We therefore add eight points to our scale to give it a 20-point index. These supplemental factors are only given scores of "2" or "0" and are operationalized according to this scale. Issue management entails how issues and their stakes have been handled in the past: either peacefully or with force. A score of "2" is added if past interactions involved force, "0" otherwise. The Arctic issue has reemerged on the global agenda with the proliferation of the melting ice. The region was once geopolitically, but not economically, important to the two superpowers during the Cold War. It was a shortcut for airplanes and submarines to deploy their nuclear arsenal in the face of World War III. However, with the fall of the Soviet Union and the technological advances with intercontinental ballistic missiles (ICBMs), the Arctic has waned as an important strategic military zone. Global warming and its consequences and opportunities in the Arctic are a relatively new phenomenon, thus past Arctic issue management is fundamentally different from contemporary and future management.

Again, past interactions only count the number of militarized interstate disputes (MIDs) for dyads occurring within a 15-year threshold.[50] The only MIDs found between these states in the past 15 years are several disputes over fishing waters between coast guards.[51] This factor's relative absence, at least in the dyads containing Russia, means that the salience of issues between the Arctic states will not be very high, which helps in conflict mitigation. Scores of "2" are given to the pair of states if there have been recent MIDs, "0" otherwise.

For the institutional context measure, the treaties and institutions for the purposes of Arctic territorial disputes are the UNCLOS, the CLCS, and the Arctic Council. The UNCLOS states the agreed-upon international legal framework that all parties have vowed to follow, with the exception of the US. The CLCS is the institution that the Arctic states have allowed to settle any outstanding territorial disputes in the Arctic. Before this commission makes its decisions, the Arctic Five have agreed to settle their disputes via the Ilulissat Declaration of 2008 in the Arctic Council. A score of "0" is given if the pair of states is a signatory to all three institutions/regimes on each side, "2" if the pair contains a state that is not a signatory to all three. The logic behind this scoring is if a state is not a signatory to all three institutions or regimes, a state that is will look at the other side with mistrust and perhaps believe that the state may renege on an agreement, as it is not legally bound to adhere to agreements. The United States, for example, gets a "0" for this measure, as it has not signed on to UNCLOS. We therefore assume that the other four countries will engage the US diplomatically with some reservations, as the US is not bound to its commitments in the Arctic region, especially when it comes to territorial claims. Disputes are thus more likely to arise, which hinders cooperation.

A score of "2" is given if there is the presence of a mixed dyad (democratic and non-democratic), "0" if it is a democratic pair of states. With this altered salience index score of a possible high of 20, our salience scale is as follows: scores ranging from 0 to 7 are considered to be of low salience, those ranging from 8 to 13 are considered to be of moderate salience to the pair of states, and scores ranging from 14 to 20 are considered to be highly salient. To come up with our scores, we use news articles, scholarly articles, and official government and IO websites to decide how salient each issue is to each state.

Russia and the United States

As Table 7.1 indicates, the disputes over territory in the Arctic are not particularly salient. The dyads involving Russia, which automatically get a mixed dyad score of two, should have the most salience and thus increased probability of the willingness to forcefully assert claims. The highest salience score is eight, and this score is shared between two pairs of states: the US and Russia and the US and Canada. Eight is a moderate salience score, and it is the minimum score for this category. The first pair of states contains the two major powers of the Arctic as well as the only pair of states embroiled in a rivalry. The United States and Russia are no strangers to each other when it comes to contending over issues around the globe. A 50-year geopolitical superpower rivalry along with a regional issue rivalry in the post-Cold War era gives the Americans and Russians the proper context to dispute access to the Arctic. The problem for this line of logic is that Russia no longer has the interest or the abilities to challenge the United States outside the territory of the former Soviet Union, at least above the level of simple diplomatic disagreements.

The US and Russia are in a territorial dispute over fishing waters that dates back to the Cold War. There is really no cause for alarm, as Mitchell and Prins find that disputes over fishing stocks are usually handled and resolved peacefully.[52] However, it still remains an unsettled territorial

Table 7.1 Dyadic salience scores for Arctic issues (Low 0–7, Moderate 8–13, High 14–20)

Dyad	Pop.	Res.	Strat.	Kinsm.	Homel./ Dep.	Mainl./ Offsh.	Issue Mgmt	Past Interact	Institut.	Advers.	Total
Russia-US	0	2	2	0	0	0	0	0	2	2	8
US-Canada	0	2	2	0	0	0	0	2	2	0	8
Russia-Norway	0	2	2	0	1	0	0	0	0	2	7
Russia-Denmark	0	2	2	0	0	0	0	0	0	2	6
Canada-Denmark	0	2	2	0	1	0	0	0	0	0	5
Russia-Canada	0	2	0	0	0	0	0	0	0	2	4
Canada-Norway	0	0	2	0	0	0	0	0	0	0	2
US-Norway	0	0	0	0	0	0	0	0	2	0	2
US-Denmark	0	0	0	0	0	0	0	0	2	0	2
Norway-Denmark	0	0	0	0	0	0	0	0	0	0	0

dispute and, therefore, the pair of states is given a score of two for the resources category. This dispute will continue, but it will be peacefully managed or resolved with clear demarcated boundaries. Therefore, the chances of the US and Russia involving themselves in an Arctic conflict is as remote as the region itself.

The next issue that gives this dyad salience is the disagreement over sovereignty for the Northwest Passage and the Northern Sea Route. Russia's coastline borders the Northern Sea Route, thus the monitoring and regulation of this route is part of Russia's sovereign right. Canada, Russia believes, has this same sovereign right for the Northwest Passage. One of the main reasons the United States takes this stance against the Russian position is its desire to utilize the Northwest Passage, which Canada sees as its sovereign territory, and not its desire to use Russia's Northern Sea Route. The Northwest Passage would shorten the shipping times between the west and east coasts as well as shorten the trips between the east coast and Asia and the west coast and Europe. The US has little to no utility for Russia's Northern Sea Route. Therefore, this dispute over the sovereignty of the emerging sea route is primarily with Canada, not Russia.

The final issue Russia has with the US in the Arctic is its continuing failure to ratify UNCLOS.[53] However, both Presidents Bush and Obama have vowed to abide by the rules of the convention, regardless of the lack of senatorial ratification. The salience of the Arctic issues between these adversaries is moderate, and if these two states were to come to blows, it will not be due to disagreement in the Far North.

Russia and Norway

Russia and Norway have the second highest salience score in the scale at seven. This score is in the low salience range. The dyad contains a territorial dispute, particularly over resource rights. Russia is claiming the same underwater territory off Norway's Svalbard Islands, where there is potential for reserves of hydrocarbons. Russia claims the underwater territory as an extension of its continental shelf and has submitted this claim to CLCS, whereas Norway claims the islands as its sovereign and suggests Russia claims territory that is exclusively part of Norway.[54] It is doubtful that this case will follow the typical path of a territorial dispute leading to a conflict spiral containing arms races, usage of powerful allies, and domestic hardliners, and an eventual Russo-Norwegian war. This is because both states rely on institutions, and the salience of the dispute is low. Russia and Norway have agreed to adhere to the decisions of the CLCS and the laws set forth in UNCLOS, as well as to address disputes within the Arctic Council.[55]

There are ethnic Russians and Norwegians living on these islands, although the settlements are not permanent and are only mining communities.[56] Therefore, this dispute did not meet the population requirement that may make this territorial disagreement more salient. Russia and Norway do not see eye to eye on Russia's sovereign claims to the Northern Sea Route, which gives this dyad a score of two in the strategic category. Norway envisions the route as an opportunity to export goods more cheaply and efficiently to one of its major trading partners, China. It wants to see an open northern passage to its Asian customer, without interference and regulation by Russia. This point of contention, thus far, has no mechanism that will decide the sovereign status of the new sea route; therefore, the contentious point may remain for some time. However, with the continental shelf dispute soon to be resolved, the salience of Arctic issues between Norway and Russia will only get lower, and not escalate. Norway's claims that the Svalbard Islands are part of the Norwegian homeland along with the mixed dyad score presently gives this pair of states a salience score of seven, which is of low salience.[57]

Since 2010, Russia and Norway have resolved a longstanding maritime border dispute in the Barents Sea.[58] Russian and Norwegian fishermen now know where they can and cannot fish. This dispute-ending agreement is to be followed by an agreement on the status of the territorial borders of the Norwegian and continental shelves. Although salience is moderate presently, it could soon be low between the states.

Russia and Denmark

Denmark and Russia disagree on Russia's CLCS claim of the Lomonosov Ridge, as the Danish assert that parts of Russia's claim are the rightful continental shelf extension of Greenland. Along with the stance Denmark takes on opening sea lanes and the mixed dyad score, this pair receives an issue salience score of six. This territorial dispute is due to the potential for underwater hydrocarbons. However, Denmark has not ruled out giving Greenland its independence, and with this possibility it is apparent that this dispute with Russia may die or be transferred to an independent Greenland before the CLCS decides on the territorial borders between Greenland and Russia. This also indicates that the Arctic issue in general is not very important to the Scandinavian state, and if Denmark drops out of the Arctic race, the dispute will be inherited by an independent Greenland. Greenland may find this matter more important than its former overseer, but this remains to be seen. If Denmark retains control of Greenland for the foreseeable future, the only dispute it will have with Russia will be over the status of the Northern Sea Route.

As Danish trade lies more to the west than to the east, the dispute over sea lane status will more likely be with Canada.[59]

Russia and Canada

The lowest salience score for a pair of states containing Russia is Canada and Russia, at four. The Canadians and Russians are in dispute over the Lomonosov Ridge, where Canada asserts that Russia's claims are parts of the continental shelves of several of its northern islands. This dispute, as with other pairs of states, should be resolved by UNCLOS laws and CLCS rulings, according to the stated commitments of the Arctic Five. Where there is agreement and what might grow in salience over time is Canada and Russia's rights to exclusive sovereignty for the Northwest Passage and Northern Sea Route. These two states believe that it will be their military, coast guards, and scientific communities that will be monitoring, defending, and researching these routes. Therefore, Canada and Russia claim, and quite forcefully through aggressive rhetoric and statements, that they have the right to regulate and tax ships coming through their sovereign waters.[60] Most of the rest of the world are taking the stance that the sea lanes are international and are not subject to regulation or taxation. This is the issue in the Arctic that may become more salient and pit Canada against its traditional NATO allies. It may also lead to a peculiar coalition with the Russians. For now, however, this point of contention has yet to heat up. If it does, it is purely an economic rights issue rather than a traditional territorial claim. Likely, if Canada and Russia are providing a service that includes coast guard monitoring and protection, port service, and navigation information, other states and companies will pay for the information and support.

United States and Canada

A surprising finding in this analysis is that the United States and Canada receive the same salience score of eight that the US-Russia dyad receives. These traditional allies and neighbors do not see eye to eye on a number of things in the Far North. The Canadians and Americans are at odds over fishing waters in the Arctic, especially over border demarcation in the Beaufort Sea. Fishing disputes have happened so often between these states that the dyad has met the requirements for a rivalry, according to dataset coding rules.[61] This pair of states receives a score of two in the past interaction category, as these types of disputes are considered MIDs.[62] The possibility of conflict escalation between the US and Canada is remote; however, these disputes have soured diplomatic relations between the states over the years and have been a source of contention.

As we have mentioned previously, fishing disputes rarely ever rise to the level of violence because they are easily settled because they are divisible stakes.[63]

Along with fishing disputes that have led to disagreements, the US dissents from Canada over the sovereign status of Canada's Northwest Passage. The US is of the opinion that the passage is international waters and can be used by American ships without Canadian permission. This does not sit well with Canada, and it remains to be seen how this issue will be resolved in the future. What is also particularly interesting is the fact that Russia and Canada agree with the sovereignty argument of these new sea lanes, against the pro-international waters proponents of the US, Norway, and Denmark, and China, Japan, and most of the EU.[64]

Canada and Denmark

The pair of Canada and Denmark receives a salience score of five. This is also unusually high for a democratic dyad. These states have territorial disputes over desolate islands surrounded by resource-rich waters. Canada and Denmark are in a dispute over the possession of Hans Island, which is basically a collection of rocks, yet has large stocks of fish and possible hydrocarbons in the Nares Strait.[65] Furthermore, the two states are at odds about the sovereign status of the Northwest Passage and Northern Sea routes. This gives the dyad a score of two in the strategic category. Lastly, Canada sees the islands close to Greenland as its territorial homeland, while Greenland is merely a dependent of Denmark. However, because the two states are NATO allies, are democracies, have both signed on to the stipulations of UNCLOS, and have agreed to settle disputes via the Arctic Council, there is a low probability for rash action by either side.

Remaining Pairs of States

The remaining pairs of states either have salience scores of two or zero. Canada and Norway's only disagreements lie with the sovereign status of the new sea lanes, and this pair of states gets the score of two. The US-Denmark and US-Norway pairs receive scores of one, and Norway and Denmark dyad are given a score of zero. The Danes and Norwegians do not share territorial borders and have no disputes over sovereign rights to resources. They also agree on the sovereign status of the opening sea lanes, have not had an MID since 1969 (over fishing rights), and are both democracies.[66] The US is not a signatory to UNCLOS and thus is given a score of two when paired with the Scandinavian states. They agree on the sovereign status of the new sea lanes, which pits them against Canada and Russia, and that may have implications in the future.

This section has demonstrated that the stakes under contention in the Arctic issue area give this issue an overall low salience score. The highest salience score is eight out of a possible 20; therefore, the salience of the Arctic to the Arctic Five is moderate to low. This allows us to assert that disputes over territory in the Arctic, whether over resource rights, sea lanes, or fishing waters, should be settled peacefully through third-party or institutional mitigation because the salience scores are moderate at best.[67] The US-Russian rivalry is a regional rivalry and not currently active in the Arctic region. Their disagreements over fishing waters have been going on for decades. US ships are more likely to use the Northwest Passage through claimed Canadian waters over the Russian Northern Sea Route. Russia's territorial claims span almost half the Arctic when sea boundaries are accounted for, yet they have submitted these claims to the UNCLOS's CLCS, which should grant most of Russia's claims if they are accurate. The Canadian-Danish dispute should be mitigated through CLCS and should be further mitigated because the two states are democracies. The dispute that the US has with Canada over the Northwest Passage should be settled peacefully with tit-for-tat negotiations. Norway's dispute with Russia will also be decided by CLCS, and there is no sign that the Russians are willing to fight with this NATO member over a relatively small area of undiscovered hydrocarbons. However, we must look at the other three measures for issues of the Arctic before coming to more concrete conclusions.

Location and Timing

Where the region of territorial disputes is located as well as the timing of the issue on the global agenda relative to other issues is relevant to the issues at stake according to our hypothesis. The importance of a region to a state is based on the distance of the capital from the major seaport of the region.[68] Another assertion of our regional measure is the number of seaports a state has north of the Arctic Circle. Finally, we argue that the timing of the Arctic issues relative to other global issues is poor, and the issue has not reached a critical level. Natural resources such as oil and natural gas have yet to become scarce in other more accessible regions. Therefore, the extraction of these resources in the Arctic has yet to become overly important to states. The Panama and Suez Canal routes have yet to become so costly that the only choice is to use the alternative Northern Sea Route and Northwest Passage. Fish is still available in more hospitable waters, and the Arctic is not an ocean of last resort for the world's fish supplies.

In the first subsection we look at the distance from the capital city of each state from each state's corresponding northernmost port and the accessibility based on the number of ports north of the Arctic Circle and available icebreaker technology the state is endowed with to assert that Russia finds the Arctic region more important than the other four Arctic coast-bearing states. The second subsection looks at how the timing of the Arctic issues relative to others on the global agenda is poor, as the states of the Arctic Five have more pressing issues elsewhere around the globe. Implications of these findings will conclude the section.

Location

Russia's Arctic concerns may be more important than those of the other states of the region. The entire northern coastline of Russia borders the Arctic Ocean. Russia has over 15,000 miles (24,200 km) of coastline bordering the Arctic Ocean.[69] This is second to Canada, which has the longest Arctic coastline because of its numerous large offshore islands in the Far North.[70] Russia, however, uses its Arctic ports and sea lanes more often than Canada. Rising global temperatures are melting the Russian permafrost, making more land arable and, more importantly, available for hydrocarbon extraction. As Russia relies on hydrocarbon exports for much of its wealth, melting permafrost will only endow Russia with more opportunities for energy export-led growth. As mentioned earlier in this chapter, the melting icepack in the claimed Russian territorial waters is projected to have nearly half of the available underwater hydrocarbons in the Arctic Ocean.[71] More offshore drilling and exploration due to sea ice melt endows Russia with more resources and more wealth. The Arctic region and its governance, therefore, are important to Russia.

Table 7.2 indicates that the capital of Moscow is relatively close to Arctic waters when compared to the other Arctic states. Murmansk is 932 miles from Moscow and is an important port for Russian exportation of resources extracted and processed in the Russian north. It also serves as an important military outpost for the Russian navy.[72] The port of Archangel is also close to the population centers of St. Petersburg and Moscow, and this port serves as the major importing center for goods entering Russia.[73] Russia uses the Arctic Ocean as a means of trade and defense in the north, arguably far more frequently than any other state on the planet.

Table 7.2 also shows that Russia has the most operational ports north of the Arctic Circle. There are five ports along the Russian Arctic

Table 7.2 Arctic regional factors of the Arctic Five

Country	Distance (Capital to Northernmost Port)	Number of Arctic Ports (North of Arctic Circle)	Number of Icebreakers
Russia	Moscow to Murmansk 923 miles (1485 km)	5	29
Canada	Ottawa to Churchill, Manitoba 1202 miles (1934km)	1	18
Norway	Oslo to Nordkapp (North Cape) 1295 miles (2084 km)	2	1
United States	Washington to Barrow, Alaska 3480 miles (5601 km)	2	2
Denmark	Copenhagen to Sisimiut, Greenland 4643 miles (7472 km)	0	3

coastline, equal to the number of ports for the other four Arctic states combined. Russia uses the Northern Sea Route to transport domestic goods from one part of its huge landmass to another. These ports are supply refuges along its near 15,000 miles of Arctic coastline. To keep the route from freezing over, Russia uses its fleet of 29 icebreakers to clear its sea lane if iced over.

The status of the Northern Sea Route, which could serve as a safer and shorter shipping lane from Europe to Asia, is of great importance to Russia. With domestic unrest in Egypt, the perpetual Israeli-Palestinian conflict, and increased piracy around Cape Horn and South China Sea, the Suez Canal route connecting Europe and Asia is becoming ever more dangerous, making insurance rates for many shipping companies near intolerable. If the Arctic sea ice continues to melt, the Northern Sea Route would be a much safer, shorter, and ultimately cost-effective route for trade between the two continents. The route will lie almost exclusively off the Russian coast, and Russia believes that the responsibility of monitoring this route and regulating its traffic will lie with its navy and coast guard. If Russia is to bear this burden, then it believes that shipping tolls should be paid to Moscow.[74] Only Canada agrees with Russia on this particular issue due to its status as a similar power in the Western region. Many other states see the route as international waters and not subject to Russian regulation. This disagreement may make this particular issue more salient to Russia if the Suez route becomes insoluble and year-round iceless waters north of Russia's coast become the norm in the future.

Russia is already heavily invested in the fate of the Arctic region and seems to be the best-prepared state in the region. Due to its presence with Alaska, the United States has its own interests in the region. However, the state ranks 47th in population out of 50, with most of this population living far from the Arctic coastline.[75] The only people living in Alaska's far north are several Inuit tribes and those working in the oil industry for extraction and pipeline maintenance. This isolated and sparsely populated part of the United States is far removed from the Lower 48. Table 7.2 shows the distance from Washington, DC, to Barrow, Alaska, the northernmost port in America. The 3,480-mile distance between the capital and the remote port is over triple the distance of that between Moscow and Murmansk. This port, along with Prudhoe Bay, is used to supply local populations with food and medicine and to supply oil workers of the Alaska pipeline system. However, the Alaska oil supply is a small percentage of the overall reserves of the United States.[76] The US has domestic supplies of oil and natural gas in many parts of its main continental territory and also imports oil from the Middle East and South America. Therefore, the urgency to explore for more hydrocarbons in the Arctic Ocean is not evident.

Alaska's fishing waters are stocked full of many species of fish. It is also the main source of crabmeat for much of North America.[77] These waters lie mainly in the Bering Sea, where sovereignty of these waters remains relatively undisputed. Occasionally Russian and American fishermen wander into each other's waters, but disputes of these kinds go relatively unnoticed and have been occurring for decades, even in the most heated times of the Cold War. The need to expand north of the Arctic Circle into the coastal waters of the Arctic Ocean has yet to become a pressing issue, although this may change if the fish migrate north because of climate change's effects on ocean temperatures.

Alaska has two Arctic Circle ports, Prudhoe Bay and Barrow. Prudhoe Bay is used in the summer to resupply workers of the oil industry, and Barrow connects the various indigenous peoples to the rest of the country. These ports are used sparingly and are not used to supply much of the domestic population with raw materials, as the Arctic ports of Russia are used. The United States has the largest and most technologically advanced navy in the world, yet it only has two operational icebreakers. When compared to Russia's 29, we can make the assertion that the Arctic region is of little importance to the world's most powerful state. What of the neighbors to America's north?

Table 7.2 shows the relatively short distance between the Canadian capital of Ottawa and its main Arctic seaport, Churchill, Manitoba.

Similar to Russia and the Northern Sea Route, Canada's main concern lies with its sovereign rights to monitor and regulate the Northwest Passage. This route could be an alternative for Europe and Asia, Asia to the American eastern seaboard, as well as an alternative to the Panama Canal that connects the American west and east coasts. The United States is particularly interested in pushing for international-water status of this new shipping lane, as its export and import costs would be reduced significantly through this shortened polar route. Canada has not appreciated this American point of view and wishes to regulate what it perceives to be its exclusive sovereign territory to protect its fragile ecosystem and indigenous people's way of life. It fears that international, mainly American, shipping traffic will pollute and destroy its high north wonders if not properly regulated and monitored.[78] Canada seems ready for the task of protecting the waters it perceives to be its own, as it has a fleet of 18 icebreakers to monitor and defend its sovereignty, including its Arctic waters, from international incursions.[79]

Also of importance to Canada is the potential for hydrocarbon extraction in its offshore Arctic waters. Canada already has large reserves of oil and natural gas, as well as the potential for huge amounts of shale oil resources in its southern mainland.[80] The need for more expensive offshore drilling exploration is a concern of the future. The Arctic issue for Canada lies with its sovereignty over the Northwest Passage, which for now is only speculation for the future. It only has one seaport north of the Arctic Circle, and there are plans to build another.

Looking at Table 7.2, the northernmost port of Nordkapp (North Cape) is a long 1,295 miles from the capital of Oslo. Most of the Norwegian population is located in the south around Oslo and the Baltic Sea. The far reaches of the Norwegian north are of little concern to many Norwegians. It has two Arctic Circle ports and one operational icebreaker. Norway is a top exporter of hydrocarbons to many European states. It has large amounts of oil and natural gas and is second only to Russia in energy exports to Europe.[81] Its concerns regarding the Arctic, therefore, lie with expanding its hydrocarbon reserves and exploration opportunities. The dispute with Russia over the Svalbard Islands continental shelf is why the Arctic is of some importance to Oslo. However, like Canada, Norway is at a satisfactory and profitable level with its current supplies. It has also vowed to respect the decisions of CLCS in its territorial dispute with Russia. Norway is more concerned with the regional well-being of Europe than with the fate of the Arctic.

Copenhagen, the capital of Denmark, is 4,643 miles from its northernmost port in Greenland. It has no ports north of the Arctic Circle. Its three icebreakers are small and utilized around the homeland to keep its Baltic Sea ports ice-free. Denmark is the sovereign protector of Greenland, which gives it the right to territorial claims in the Arctic. It has suggested that it will not rule out granting Greenland its statehood and autonomy in the near future.[82] Greenland has valuable fishing waters and a possibility for hydrocarbon reserves. However, with the Danish suggestion that it might be granting Greenland its sovereignty, the Arctic region is a minor concern to Denmark, even less so than to their Norwegian neighbors.

The location measure presented in this section has shown that the Arctic region is of more importance to Russia than to any other Arctic state. Russia's capital is closest to its northernmost port, it has the most ports north of the Arctic Circle, and it has the world's largest icebreaking fleet. It uses its Arctic coastal waters and ports to transports goods and raw materials to different parts of its vast landmass while other countries have other routes. The economic implications of the region for Russia are greater and more developed than those of the Western Arctic states. Russia is presently utilizing the region and is preparing to expand its use as the sea ice recedes. It will have the ability to drill for oil and natural gas in more places offshore as the Arctic Ocean heats up. As Russia has the most to gain from Arctic warming, it also has the most to lose if it is not cooperative with the other Arctic powers.

Canada is looking forward to the economic potential of the Northwest Passage as ice in this waterway becomes more and more sparse. It has the second-largest icebreaking fleet to protect its interests in the Far North. It only has one Arctic Circle seaport, but it is preparing to build a military seaport in the northern reaches of Hudson Bay.[83] However, unlike Russia, Canada does not rely as heavily upon its Arctic waters for economic well-being and domestic transport of goods. Ninety percent of Canada's population is within 100 miles of the US border.[84] Russia has five major Arctic seaports and its second-largest city, St. Petersburg, is only 500 miles (800 km) from the Arctic Circle.[85] It trades heavily with the United States and has a reliable rail and highway system to supply its citizens. Therefore, the Arctic region is important to Canada, but nowhere near the level of importance to Russia.

The populations and capital cities of the United States, Denmark, and Norway are far removed from their Arctic regions. Their icebreaking fleets are small and utilized outside of the Arctic in most instances. Usage of the Arctic region for these states for economic well-being is not essential. With the regional measure showing how the Arctic issue area will probably not lead to discord and conflict, the next part of the section presents the issue measure of timing.

Timing

Vasquez and Mansbach present an "issue cycle" conceptualization for the global agenda.[86] They see international issues as going through "cycles," where there are four stages through the life of an issue. The timing of the Arctic issue on the global agenda is thus poor, and perhaps in the future the global agenda will be ready for its attention.

The first stage of the issue cycle is called the genesis stage. This stage happens when attention by global actors shifts from one critical issue to another.[87] The old issue of the day is usually replaced because the stakes have been distributed, and the issue, for the most part, has been settled either through violence or peaceful settlement. The genesis stage changes as one single dominant issue emerges as the critical issue.[88] The second stage in the issue cycle is called the crisis and ritualization stage.[89] It is called the crisis stage because there is an atmosphere of crisis among decision makers, there is a change in the objectives of foreign policy among states, which produces a degree of uncertainty as shifting alignments may arise, and there may be some aspect of an accepted morality that is threatened.[90] The third stage in the issue cycle is the dormancy and decision-making stage. In this stage, the issue's salience is reduced, as crisis is either averted or solved through conflict. Dormancy happens when the stakes of an issue are relegated to the periphery of public attention.[91] The final stage is called the authoritative allocation/removal from the agenda stage. In this stage, rivals are eliminated, usually through force, or states come to a mutual agreement. When these occur, the issue is for the most part resolved and removed from the global agenda. What of the Arctic issue? Where does it lie in the issue cycle, and what stage is it most likely in? How does its existence as a global issue relate to other issues on the global agenda; how has its timing affected how it is being handled?

The Arctic issue can be considered a regional issue pertinent to many great powers. However, not all policymakers and the public of these states see this issue as pressing and deserving of their time and energy. Therefore, the issue of the Arctic to some may not be as salient as to others. The Arctic issue is still in the genesis stage, as it is vying for status of critical issue alongside other issue areas, such as the global economy, terrorism, or China's rise. The issue may reach critical status, and thus crisis stage, if Manhattan floods due to rising temperatures, hurricanes in the Gulf of Mexico become more frequent, or the global supplies of oil and natural gas become dangerously scarce. Until then, the Arctic issue will remain in its genesis stage until its particular stakes become more salient to the states involved.

The Arctic issue will then move into the decision-making and authoritative allocation/removal from the agenda stages rather easily, as when global warming and climate change become a reality and thus more accepted. The need for oil and gas, faster shipping lanes, and seafood should someday be considered more important than the fate of the polar bear, the Inuit people, or the scenic beauty that once dominated the region. The Arctic may someday reach its critical status, but until then it will remain in its genesis stage and will be handled cooperatively by the states with stakes in the region. The regional measure shows that Russia finds the Arctic region essential to its economic growth and well-being, and not so much for the other four states. With submissions of claims to its Arctic territory to the CLCS, Russia is hoping for a peaceful endorsement of its continued exclusive use of its Arctic waters. As it has a very good case to get what it wants, Russia should be able to continue its ambitions unabated. Furthermore, what the public of these Arctic states think about the Arctic issue may supplement this view, which brings us to our final issue measure, public opinion.

Public Opinion

Hutchinson and Gibler present results that suggest the salience of territorial issues for domestic audiences is greater than salience for other international issues.[92] Furthermore, Gibler, Hutchinson, and Miller find evidence for individual-level feelings of high salience for territorial issues.[93] In short, the public cares more about territorial issues than other international concerns. Homeland territorial issues are often more critical to any sort of boundary or international issue on the agenda. The Arctic territories of the five states under examination are not considered part of the homeland and are thus not as salient to the public and elites of these states.[94] Therefore, the disputes over territory in the Arctic are atypical and should not fit the dispute pattern of a high salient issue to the public.[95] We examine the Arctic issue with the latest public opinion polls taken in the five border states of the polar region. We find that the public of these states do not consider the Arctic particularly salient. Issues in the Arctic should not lead to a conflict spiral because the salience of these disputes, some territorial, is low due to the location dynamics of the issues and long-term time horizons associated with some of the issues.

The public opinion measure asks a simple question: What do the various publics of the Arctic Five states think about the Arctic issue and the stakes involved? Policymakers are usually only attentive to what the public deems important, ergo how pressing the issue of the Arctic is to states could be directly related to how the general public, as well as some in elite circles, feel about the issue. Here we present public opinion polls of the five states of the Arctic on various questions about the Arctic issue. The most comprehensive public opinion surveys on issues of the Arctic and the stakes involved are found in the Canada Centre for Global Security Studies Munk School of Global Affairs/Walter & Duncan Gordon Foundation (CCGS hereafter) 2011 survey.[96] This poll surveys random samples of the public of the eight members of the Arctic Council for opinions on the various issues pertaining to the Arctic. Presented below are results of four polls taken with a focus on the Arctic Five.

Table 7.3 gives the attitudes of the citizens of the Arctic Five on how they want their home governments to deal with border and resource-sharing disputes in the polar region. The majority of Canadians and Russians wish to see their leaders take a firm stance on Arctic questions. These opinions are correlated with the salience measure presented earlier in this chapter. The sea route sovereignty disputes with Russia and Canada on one side, and basically the rest of the world on the other, make the people of Canada and Russia uneasy about high international shipping traffic flowing close to their homeland without compensation. Therefore, they feel more strongly about taking a firm line on the disputes arising in the Arctic. The majority of the American and Scandinavian public see Arctic disputes as workable and open to compromise. This indicates that these publics are willing to allow their home governments to come to terms with the other side if a territorial dispute arises in the Arctic. If this is the case, these compromising states are more likely to concede to Canada and Russia about their sovereign right to monitor and regulate their territorial waters, thus allowing for a peaceful settlement of this moderately salient issue.

Table 7.3 Preferred approach to resolving Arctic disputes

	Canada	Russia	United States	Norway	Denmark
Firm line	43%	34	10	8	5
Compromise	39	33	30	49	24
International Treaty	11	14	25	35	24

Source of data: CCGS Survey: January 2011.

A sizable percentage of Americans, Danes, and Norwegians also believe that the Arctic should be considered international waters, thereby not subjected to one state's territorial claims or jurisprudence. Introduced in the beginning of this chapter, the International Treaty on the Antarctic stipulates that no state can make territorial claims on the continent, as Antarctica is considered a global "commons" and is open to any country who wishes to explore it. This is also the view of many of the non-border states with Arctic interests, namely China, Japan, and the states of the EU.[97]

However, it does not seem that the same rules are able to apply to the Arctic region for several reasons: 1) the Antarctic is a landmass, while the Arctic is an ocean that is mostly frozen; therefore, different international laws apply. The UNCLOS is the major mechanism of how maritime disputes are settled, while the treaty over the Antarctic region is unique to the specific area, which is completely void of permanent residents. 2) The Arctic potentially has huge supplies of hydrocarbons and other minerals, as does Antarctica. Yet the major powers are in the northern hemisphere. Russia, for example, has ready access to the Arctic Ocean, and exploratory and extraction costs would be much lower if it stayed in its own backyard. Furthermore, the Antarctic either does not have the potential for natural resource exploration and extraction or the cost of exploratory drilling is too high.[98] 3) The Arctic region has accessible fishing waters and shortened shipping lanes. Fishing is accessible in the Antarctic, but not as close to the major population centers of the world and their markets. The same goes for southern shipping lanes; they are not strategically important to the powers looking for alternatives to the Panama and Suez Canal routes. Therefore, the international cooperation found in the Antarctic is not comparable for the Arctic.

A CCGS poll that asks about the public opinion on the territorial nature of the Northwest Passage finds that a three-fourths majority of Canadians believe these waters to be exclusively Canadian. This is because of the heightened public awareness campaigns elites and leaders have lobbied to the Canadian public.[99] Canada prides itself on being environmentally friendly and is afraid that unregulated international shipping lanes would have an adverse environmental effect on the Canadian north. Furthermore, many Canadians believe that declaring the Northwest Passage international waters would be an infringement upon Canadian sovereignty and the native Inuit people's way of life. Therefore, for Canadians, the Northwest Passage must remain Canadian.

The majority of Scandinavians believe that the Northwest Passage should be declared international waters so that ships can move freely through these waters without subjugation to Canadian authorities. This correlates with the stance of Scandinavian leadership, who also believe that this potential sea lane should not be under anyone's authority. The respondents from these countries, as well as from Canada, seem to correlate with the official Arctic policies of their governments, which follows the measurement of our public opinion measure.

Peculiarly, the majority of respondents from the two major powers of the Arctic Five, the United States and Russia, either do not know what the status of the Northwest Passage should be or did not respond to the question. This is particularly puzzling for Russia, whose government agrees with both Canadian respondents and the Canadian government that Arctic sea lanes should be considered territorial waters under the supervision of the coastal governments. Russia's stance on the Northern Sea Route is the same as Canada's on the Northwest Passage, yet the Russian public seems unaware of this similar stance and does not know what the Canadian passage's territorial status should be.

The American public is indecisive and unaware of the status of the Northwest Passage. Nearly two-thirds of Americans responded to the question by either giving no answer or stating they did not have enough knowledge to answer the question. This indicates that American elites and leaders are either not paying attention to this dispute, even though they disagree with the stance of Canada and Russia, or that the American public does not care enough about this issue to acquire knowledge about it. Whatever the reasons, the opinions of the American respondents show that the sea lane issue is not atop many Americans' foreign policy concerns, making the Arctic issue in the United States not very important among the public's foreign policy issues.

Table 7.4 shows how aware people of the Arctic states are about Arctic institutions. Public awareness of the mechanism for dispute resolution, the Arctic Council, is dismal for all Arctic Five states. This implies that home governments and elites are not informing their populace about the disputes and issues over the Arctic, or the public is relatively uninterested. The Arctic Council is the primary mechanism for Arctic negotiations and settlements, yet the majority of respondents are either vaguely aware or not aware of its existence. When the public is not aware of the primary mechanism for Arctic cooperation, it implies that the issues of the Arctic are not atop the foreign policy agendas of both leaders and the public, which allows for fewer chances for conflict and more chances for cooperation and peaceful settlement.

Table 7.5 shows public opinion of the five states for support of the Arctic Council. Large majorities from Canada, Denmark, and Norway are supportive of the organization when it is explained to them, which

Table 7.4 Awareness of Arctic Council

Respondents asked: Have you ever heard of an intergovernmental forum or group called the Arctic Council that is made up of eight countries with Arctic regions?

	Denmark	**Norway**	**Canada**	**Russia**	**United States**
Clearly	30%	20	15	7	2
Vaguely	21	20	36	14	14
No	49	60	49	79	84

Source of data: CCGS Survey: January 2011.

Table 7.5 Support for Arctic Council

Respondents asked: Do you support or not the idea of an Arctic Council so the eight Arctic nations can work together on common Arctic issues, instead of each one working independently?

	Support	**Don't Support**
Canada	82%	18%
Denmark	76%	24%
Norway	72%	28%
Russia	55%	45%
United States	56%	44%

Source of data: CCGS Survey: January 2011.

indicates that they want their governments to be accommodative to the other interested parties of the Arctic. The two major powers' respondents show nearly half support for the Arctic Council, with the other half in support of their governments taking an independent path on Arctic issues. This implies that because the overwhelming majority of respondents are in support of the Arctic Council, they do not self-identify nationalistically with their home country's Arctic claims. The issue will therefore remain less important and of little concern to the public, paving the way for peaceful settlement over conflict escalation.

Table 7.6 presents the public opinion of the states that were asked about whether or not the Arctic Council should be expanded to build peace or become a military alliance. Large majorities from all states believe that the council should expand its role for peace building. Again, wanting your state to be bound to an international organization for the promotion of peace indicates that nationalistic tendencies about the state's Arctic territorial claims are not present. This reinforces the notion that Arctic issues will more likely than not be settled peacefully.

Table 7.6 Support for expanded Arctic Council mandate

Respondents asked: Do you think the Arctic Council should also cover areas like **peace building** and **military security** in the Arctic?

	Peace building (% yes)	Military security (% yes)
Canada	82%	62%
Denmark	57%	48%
Norway	66%	51%
Russia	85%	81%
United States	69%	51%

Source of data: CCGS Survey: January 2011.

Compared to the positive public opinion about expanding the Arctic Council for peace building, there is less support for the Arctic Council serving a military security role. However, there is majority support for a military role in Canada and Russia. As Canada and Russia have nearly 90 percent of the Arctic coastline and are vying for territorial support of their opening sea lanes, their publics may be more informed about the stakes involved for their countries on Arctic issues. Another possibility is that more Canadians and Russians identify the Far North as their backyard and consider it to be in need of military protection. However, with the support the public gives to institutional mediation, it seems like this opinion on military deployment in the Arctic is simply a fear of the unknown future. Only half of Americans, Danes, and Norwegians see the need for the Arctic Council to have a military role. From whom would the Arctic military be protecting the Arctic?

Overall, public opinion of the Arctic Five states indicates that the issues of the Arctic are not of much concern to the public. The Arctic is a far removed place away from the daily lives of the five states' citizens, and many seem to have limited knowledge of the situation and the players. Although territorial disputes are involved, nationalism seems quelled because these questions are not connected to the homeland. Due to the lack of widespread public support of settling Arctic disputes forcefully, the probability of conflict is small. Leaders and elites have no reason to bring the Arctic issue to the forefront of national policy, as it does not seem important enough for majority support nor is it critical to electoral fortunes. The public opinion measures presented in this chapter indicate that peaceful settlement over Arctic disputes will most likely occur. Along with the low salience of the issue area, the remote location of the Arctic, the subjugation of the Arctic issue area to other issue areas on the global agenda, and the low awareness among the public of the

Arctic Five, we assert that the Arctic is one region where peaceful cooperation can be enfranchised.

Conclusions

The ice in the Arctic is melting at an alarming rate. It seems that with this rapid melt, many in the Western media and academic realms are predicting that the Arctic will be the next region of conflict over resources—a free-for-all over the last known untapped reserves of oil and natural gas.[100] Furthermore, the race for new fishing waters as well as rows over new shipping lanes between Russia and Canada on one side, and the United States and the rest of the world on the other, will further heat up disputes and discord in the Far North. However, using the issue-based perspective, we find that this is not the case. The issue is not salient, the region is remote and has very few actors with ports and Arctic capabilities, the timing of the issue on the global agenda is poor, and public opinion is either unaware or highly in favor of peaceful settlements of the issues.

Exploration and extraction of natural resources and the territorial rights to these activities will be solved according to UNCLOS rules through the CLCS. The territorial delimitations of new prospective fishing waters should also be solved through this convention. The prospects of new shipping lanes in the Far North may cause a rift between Russia and Canada on the one hand, and the rest of the world on the other; however, the peaceful issue linkages of the other Arctic issues should pave the way to negotiated compromise or settlement. The UNCLOS could also be a harbinger to settling this dispute. Therefore, this chapter has demonstrated that Russia can be dealt with, that it can compromise, and that coercive diplomacy is not always used by the Russian state. This issue area can therefore be informative to handling issues in other areas between the West, CIS, and Russia.

With these findings in mind, Russia has no reason to behave in a power politics fashion in the Far North. In relation to the framework presented in Chapter 2, there are no rivalries active in the region (the US-Russia rivalry is only active in the former Soviet Union), the salience for the disputes is low, the location of the disputes is distant from the homeland, and public opinion tends to support cooperative agreements for settling differences among the Arctic Five.

Russia's northern coastline makes up nearly half the land that borders the Arctic Ocean. Russia is the most prepared and most willing to actively explore and develop the Arctic. It has the world's largest icebreaker fleet. It has the most Arctic-bordering ports. Gazprom and other Russian energy conglomerates are investing capital in Arctic offshore drilling. Furthermore, Russia is the first Arctic state to sign on to the UNCLOS convention and the first to submit its territorial claims to the CLCS fact-finding organization. Every other Arctic state has followed suit, except the United States, and become a member of UNCLOS. Therefore, Russia seems ready to lead in the Arctic in a peaceful manner. These findings demonstrate that not all disputes are alike. Some have different paths, and the context and situation evident in foreign policy questions matters a great deal in determining how they are handled. The Arctic will surely be a test for the international community, but the analysis provided here demonstrates that it is likely these disputes will be handled peacefully, and Russia will not use its leverage in energy power or cyber power to try to achieve its ends.

8
Conclusion: Russian Coercive Diplomacy after the Cold War

Introduction

Russia is reemerging with new forms of power. It has seen a renaissance in its status due to a consistent political regime run by Putin and an explosion in energy revenues that coincided with Russia's state takeover of its energy conglomerates, its burgeoning cyber power, and its preparation for the future with its growing Arctic maritime power. Thus it once again has the power and capabilities to be a strong political actor, but its international reach mostly applies to its immediate regional environment. Countries of the former Soviet Union have been subjected to the coercive diplomatic tactics that many states around the globe experienced during Soviet times. Russia is attempting to reclaim political control of its former empire through coercion. Any outside state that attempts to undermine Russian influence in the region is similarly met with coercive diplomacy of some form. It seems that former Soviet states that attempt to align with other power centers, such as the American-dominated West, are the ones who bear the brunt of Russia's coercion for diplomatic subversion. Unfortunately for Russia, these tactics have not achieved the political ends it desires. Its more nuanced policy in the Arctic is paying off; however, the use of power politics in post-Soviet space only moves these states closer to the West.

This chapter reviews how looking at new forms of power through our theory of situational coercive diplomacy has worked throughout this volume and how it can be a more effective lens to study international relations and foreign policy behavior, not just for Russia, but for all states. Here we offer a new way to study foreign policy by using Russia as a focal point and hope that scholars in the future will take our viewpoint into consideration for their studies. Situational coercive diplomacy as a framework can provide a pathway to analyze, predict, and explain foreign policy action and how new forms of power are utilized.

Situational Coercive Diplomacy

Context and situation matter in foreign policy. Grand theoretical traditions have little predictive currency in international interactions because they fail to account for historical and situational variables that change according to the unit of interest. Here we focus on situational factors such as rivalry and salient issues to predict and examine when Russia will use coercive diplomacy. In the territory of its former empire, Russia is not attempting to control or reconquer the region with its military might; it is attempting to control the destiny of the region through new means of power projection.

As we have seen, Russia now uses cyber capabilities, coercive energy power, and even limited warfare to engage its rivals. To understand when and how Russia utilizes power, we must look at the international situation, the internal capacity of Russia, and domestic factors often ignored by other theories. While we cannot hope to be universally comprehensive in an examination of every factor important for foreign policy decision making, here we hope to be as complete as possible. There are directions we have left unexplored, such as culture, gender, or ethnicity, but the factors we do explore account for both Russia's successes and failures in the international realm.

Barkin notes that realism as a theory of foreign policy cannot hope to be both predictive and normative since these goals are logically inconsistent.[1] Either the world works as it does, or it does not. This work follows this advice in some ways. Power politics behaviors can be empirically observed under certain conditions, but this book is neither advocating that these behaviours should be the norm nor that they are very effective. Instead we argue that the coercive diplomacy has little long-term currency in foreign policy. The means never seem to accomplish the ends. Instead, when a state repeatedly attempts to coerce, bully, and dominate another state, the victimized state will find whatever leverage it can to counter the bully. As we have discussed in this book, Russia is often that bully, and it fails to achieve its desired ends in the long run because of this strategy.

We end this volume with an examination of how Russia engages other states in the Arctic region. The findings of Chapter 7 are likely harbingers of the future. Putin's rule must eventually end. A reevaluation of the means and objectives of Russian foreign policy is likely to come soon, just as it came during the Obama Administration after the era of the Bush Doctrine. What path Russia takes in the future really depends on the strategic culture it finds valuable. Clearly, coercive diplomacy has not succeeded in bringing the power and status Russian leadership thought possible. The problem with foreign policy is that lessons are tough to learn.[2] They tend to be generational. When objectives and methods fail, some states tend to double down on these failures and try harder, thinking that the path was right but the application failed.

Coercive foreign policy decision making has often failed. The means and the ends of foreign policy must be connected. As Vasquez notes, few scholars actually examine the outcomes of foreign policy goals.[3] Instead, they tend to look at each situation through the lens of the tactic instead of the outcome. Successes only lead to the reinforced beliefs about the foreign policy path chosen. The problem is that failures occur more often than successes, but these failures are ignored because they do not fit in with the worldview of the actor. Only time will tell if Russia will shift course and act in a different manner. As long as Putin and his immediate successors remain involved, it will likely stay the course. How Russia acts in the Arctic and its restrained use of cyber power (Chapter 4) do give us hope for the future.

What Do We Know about Russian Foreign Policy Now?

This project has shown that coercive diplomacy and the use of new power politics are at play in Russia's foreign policy apparatus. However, the global reach of historic Russia has been muted by a reduction in overall military and economic power when compared to the superpower status of the Soviet Union. Russia cannot challenge the United States on the seven continents nor on the high seas. It can only assert its interests in its regional sphere of influence. Attempts at making inroads in Europe and Asia are met with coordinated Western power, and countering US interests in the Middle East has also been met with the diplomatic condemnation for failing to act in Syria to protect civilians.[4] In the Arctic, Russia makes demands but has remained restrained in its actual actions.

Chapter 3 showed that Russia is still the United States' principal rival.[5] Enemy perceptions by elites and the public remain in both states. Although the rivalry is no longer a geopolitical battle for global influence, it is alive and well, as disputes over issues are plenty in post-Soviet space. The rivalry is thus over certain global issues and issues pertaining to the region that Russia feels is its exclusive sphere of influence and where Western influence is unacceptable. Global issues where the US and Russia do not see eye to eye lie in NATO enlargement, American anti-ballistic missile plans, Russian arms sales to rogue states, and the recent uprisings in the Arab world. Regional issues involve Russia's pushback against American and Western influence in post-Soviet space, made possible by the weakened Russian state in the 1990s and exemplified by actions in Crimea. Chapter 3 also uncovered the domestic sources of rivalry and issue salience by utilizing public opinion polls as a tool of measurement. Issue salience is found to be higher when public opinion on a particular issue is closer to unanimous rather than divided. Majority negative opinions about Russia's foes are found to be highest for the two states Russia is embroiled in a rivalry with: the United States and Georgia. There are also high percentage opinions about the opposition to NATO enlargement as well as the suspicions about its motives and purpose in the post-Cold War era. Furthermore, high issue salience was found via the public opinion of Russians over specific military interventions: conflicts involving Russia's fellow Slavs, the Serbians. Russia took coercive and hardliner stances against the United States and NATO over the interventions in Bosnia in 1992 and Kosovo in 1999. Russian and American troops have not been in such proximity to each other since the days of East and West Berlin. Looking at domestic sources for the purpose of measuring foreign policy actions is an important contribution to this overall effort. While public opinion alone cannot drive foreign policy, it does play a role in either shaping elite opinions or funneling elite objectives.

Chapter 4 found that Russia is potentially the most dangerous state in cyberspace. We therefore expected Russia to use these capabilities in a power politics fashion towards any state that dares challenge or offend it. However, Russia has only used its cyber capabilities against three states, two of which are rivals and the other is part of the former Soviet empire. Estonia was hit with a two-week barrage of vandalism and DDoS-type cyber attacks in reaction to the removal of a Soviet war memorial from the central square of the Estonian capital, Tallinn. Georgia suffered the same types of attacks in tandem with the August 2008 conflict it fought with Russia. The United States and Russia have exchanged relatively benign cyber attacks, and we attribute this to a supplemental strategy of a typical rivalry relationship where each side is attempting to disrupt the other by any means possible. Furthermore, when compared to other states that have cyber capabilities and have engaged in cyber attacks, Russia has used its awesome potential very sparingly. China and the United States have engaged in nearly three times the cyber attacks as Russia. Israel, the Koreas, India, and Pakistan have also been more active in cyberspace than Russia.[6] As Russia no longer has a global reach, at least compared to the United States and China, its cyber interactions remain local and with their longstanding American enemy. Other more active cyber powers are either embroiled in more rivalries or are in more pressing regional disagreements. In short, although Russia has the ability to use its cyber power to wreak havoc on its enemies, it prefers more traditional power politics tactics to achieve its foreign policy goals, as demonstrated by actions in Ukraine.

Chapter 5 examined Russia's peculiar coercive energy policy in its former empire. Economic statecraft is utilized in this area because Russia attempts to rein in the governments of its former empire by rewarding or punishing states with adjusting the price they pay for Russian natural gas. Ukraine, Georgia, and Moldova are examples of Russia's coercive energy policy, where they pay more for natural gas as they continually attempt to escape Moscow's grasp and align themselves with the West. Belarus, Kazakhstan, and Armenia are states that are rewarded with lower relative natural gas prices for their continued friendly diplomatic relations with Russia. Furthermore, states such as Ukraine, which provides important pipeline transport to Europe, should also see lower prices for natural gas. The more Russia punishes these states with higher gas prices, the further these states veer away from complying with the goals of Russian foreign policy. Indeed, it seems that Russia is building pipelines to circumvent transit countries in post-Soviet space while at the same time many former USSR states have attempted to diversify their energy suppliers.

Chapter 6 compared two regions of post-Soviet space: the Caucasus and Central Asia. We find that the Caucasus region has high energy salience, with the presence of the United States, Russia's principal rival, in the region's energy issues. Past animosities, the fact that this region serves as a transit route for non-Russian gas, and membership in pro-Western organizations such as GUAM and the EU Neighborhood Policy makes this region salient to the Russian foreign policy elite. This has led to high coercion of the states of this region, particularly Azerbaijan and Georgia, over energy issues. Furthermore, Russian public opinion has been clear in its approval of the use of coercive energy tactics on the states of this region and the major power involved in its energy affairs, the United States. Central Asia, on the other hand, is found to have low energy salience. Therefore, the evidence of the four impact factors correlate with the low energy coercion by Gazprom and the Russian state in this region of post-Soviet space. These states belong to the pro-Russian CSTO, Shanghai Cooperation Organization, and the Eurasian Economic Community that keep the salience for these states low. Furthermore, there are no past animosities, few ethnic Russians, no pipelines that circumvent Russian territory to Western markets, and most leaders of these governments are authoritarian and like-minded with the foreign policy decision of the Russian state. All of these factors are why the salience of Central Asia is low for Russia.

In Chapter 7 we argued that in a particular issue area, the Arctic, cooperation with just about any actor is possible. The presence of international norms, institutions, and the use of international law provide a blueprint for cooperation over conflict. This avenue is possible when dividing the potential spoils that receding ice in the Arctic Ocean will bring. The low salience of the issue, even with the presence of territorial disputes, the remote location of the Arctic, the poor timing of the issue relative to others on the global agenda, and the indifferent views found in public opinion polls about the Arctic issue all contribute to the probable cooperative path that Russia will follow with the West. Natural resource exploration and extraction, governance of the opening shipping lanes in the Far North, and fishing waters will all be decided in meeting rooms rather than on the battlefield. With this promising finding, where Russia does not use power politics strategies and can be dealt with through negotiations, we are overall optimistic about the future course of Russia foreign policy, as long as it continues to act the way it does in the Arctic in other global interactions and the US does not antagonize Russia further.

Path Forward for Russian Foreign Policy Analysis

The path forward for Russian foreign policy analysis, and just about every state, is to first look at the sources of the foreign policy goals and then the outcomes. If coercive diplomacy is utilized and it also fails, the situations that bring about coercive diplomacy need to be acknowledged and reconsidered. States facing a strong public opinion, salient issues, and rivals cannot afford to ignore these threats and demands. At the same time, they cannot afford to use the same failing methods over and over again.

Russia's use of coercive energy policy illustrates this point. By charging higher prices for natural gas or pipeline usage in order to get states to fall into line and accept Russian dominance, these states often accelerate their alignment with the West. Yet Russia continues to use this tactic regardless of the opposite outcomes of the intent of the tactic. Why? It seems that coercion is what Russia knows and that accommodation in the former Soviet Union is something that Russia is only capable of achieving when the issues on the table are of low salience. As it tries to regain its former empire by means of regional dominance,

Russia is actually losing ground by continuing the power politics path. States such as the three Baltic states have fully integrated with the West, recently attaining NATO and EU membership. Ukraine has shifted its preferences between East and West, but as Russia continues its coercive energy policy, even perceived allies of Putin in Ukraine are growing tired of this strategy.[7] Georgia has suffered a military engagement with its former master, and its Western alignment ambitions are stronger than ever. In this context, Russia looks at a familiar scapegoat to blame for this insolence: its longtime rival, the United States.

The United States and Russia have been dealing with each other as enemies for over 60 years. The two states vied for dominant global influence over political and economic systems and fought several proxy wars to contain each other's influence from spreading. With the fall of the Soviet Union, this rivalrous relationship has shifted from a global geopolitical rivalry to a regional rivalry over issues. Russia and the US still mistrust and disdain each other, and this has prevented the rivalry from terminating. Negative public perceptions of each other are still ongoing, and until these perceptions are changed, the rivalry will continue. Russia and the US disagree on issues including weapons sales, alliances, post-Soviet space, regime change in the Arab world, and military interventions. However, these two adversaries have yet to escalate tensions in cyberspace, as both have the potential to hurt each other digitally. We attribute this to deterrence dynamics and the fact that cyberwar would fall out of the normal relations range.[8] Cyberwar is a new tactic, and old enemies using new tactics with each other could lead to an unknown future.

The situation in the Arctic gives hope for the termination of this longstanding rivalry. Putting aside the potential environmental catastrophe that permanent Arctic Ocean ice melting may bring, states are making claims to the potential spoils made available due to this warming trend. Russia, the United States, and the three other states bordering the Arctic Ocean—Canada, Denmark, and Norway—have set up an international regime in the Arctic Council. Here these states have laid out their claims pertaining to underwater territory with potential hydrocarbon reserves and new fishing waters. There is also potential for new shortened shipping lanes due to ice melt. With these territorial disputes and the Arctic's enormous economic potential, one would think that the region will be the next hotbed for conflict. However, the Arctic Council has brought these states together and has ensured that the precedents of international law are followed, so all may benefit and conflict is avoided. UNCLOS rules, which delimit underwater territory and excusive economic zones for states, should settle all of the territorial disputes. So far, these organizations and agreements have prevented conflict in the region. Hopefully, this avenue of cooperation remains open.

International regimes, institutions, and law have brought two long-standing enemies to the negotiating table over territorial disputes with great economic potential. This shows that there are avenues out of rivalry relationships and that coercive diplomatic tactics are not always necessary, even between the United States and Russia. The tricky part is getting states such as the US and Russia to learn from their cooperation in the Arctic and apply it to other regions and other issues. Learning, especially when on the long road out of a rivalry, can take time. However, it is not impossible. We hope that this book gives the readers a glimmer of hope in a world where hope seems lost. If Russia and the United States can find other strategies besides power politics tactics, and take the salience of the issue in dispute into consideration, with public opinion as an indicator, perhaps conflict escalation can be avoided, and we can find more paths towards peace.

Notes

Chapter 1

1. Lee, Carol E. and Dion Nissenbaum. 2014. "US Blames Rebels for Missile Attack on Malaysia Airlines Jet in Ukraine." *The Wall Street Journal* 7/18/2014. Accessed 7/28/2014, available at: http://online.wsj.com/articles/u-s-blames-attack-on-rebels-1405729164
2. Chapman, James and Rosie Taylor. 2014. "David Cameron Likens Russia's Actions in Ukraine to the Aggression against Belgium and Poland that Sparked the World Wars." *Daily Mail*, 7/31/2014. Accessed 8/3/2014, available at: http://www.dailymail.co.uk/news/article-2711452/David-Cameron-likens-Russia-s-actions-Ukraine-aggression-against-Belgium-Poland-sparked-world-wars.html
3. DeYoung, Karen. 2014. "Obama Says Malaysian Plane Shot Down by Missile from Rebel-held Part of Ukraine." *The Washington Post*, 7/18/2014. Accessed 8/1/2014, available at: http://www.washingtonpost.com/world/missile-downs-malaysia-airlines-plane-over-ukraine-killing-298-kiev-blames-rebels/2014/07/18/d30205c8-0e4a-11e4-8c9a-923ecc0c7d23_story.html
4. Tran, Mark and Alan Yuhas. 2014. "US Intelligence: Rebels Likely Shot Down MH17 'by Mistake'—As It Happened." 7/22/2014, Accessed 7/28/2014, available at: http://www.theguardian.com/world/2014/jul/22/mh17-eu-foreign-ministers-mh17-sanctions-russia-live-updates
5. Lee and Nissenbaum 2014.
6. Economic sanctions imposed by the United States and the European Union include the freezing of assets of Russian leaders, travel bans for these individuals, and as a result of the Malaysian Airlines disaster, newer sanctions have been targeting Russia's energy sector, which is the foundation of Russia's economy.
   Baker, Peter, Alan Cowell and James Kanter. 2014. "Coordinated Sanctions Aim at Russia's Ability to Tap Its Oil Reserves." *The New York Times*, 7/29/2014. Accessed 8/2/2014, available at: http://www.nytimes.com/2014/07/30/world/europe/european-sanctions-russia.html?_r=0
7. Easton, David. 1953. *The Political System: An Inquiry into the State of Political Science* (New York: Alfred A. Knopf).
8. Nye, Joseph S., Jr. 2004. *Soft Power: The Means to Success in World Politics* (Cambridge, MA: Perseus Book Group).
9. Some use the term "Near Abroad" to describe these territories. We do note that the term "Near Abroad" is a Russo-centric term where many of the people of the former Soviet Union find the term to be somewhat pejorative and condescending. We try to avoid it where possible.

10 Mordorovets-Getty, Sasha. 2013. "Ukraine, Russia cut $15B Gas Deal." *Al Jazeera America* 12/17/2013. Accessed 12/20/2013, available at: http://america.aljazeera.com/articles/2013/12/17/russia-restores-oilflowstoukraine.html
11 Nygren, Bertil. 2008. *The Rebuilding of Greater Russia: Putin's Foreign Policy towards the CIS Countries.* (New York: Routledge).
12 Nygren 2008. The Soviet Union had a population of roughly 287 million on the eve of its collapse, and today's Russian population is roughly 144 million.
13 Lally, Kathy. 2013. "Russia Tries to Improve Life Expectancy with Laws Curbing Drinking, Smoking." *The Washington Post* 2/21/2013. Accessed 7/15/2013, available at:http://articles.washingtonpost.com/2013-02-21/world/37208917_1_life-expectancy-infant-mortality-rate-president-vladimir-putin.
14 Lally 2013.
15 Putnam, Robert D. 1988. "Diplomacy and Domestic Politics: The Logic of Two-Level Games." *International Organization* 42 (3): 427–460.
16 Valeriano, Brandon. 2012. *Becoming Rivals: The Process of Interstate Rivalry Development* (London: Routledge).
17 Kessler, Glenn. 2012. "Mitt Romney and Russia: Geopolitical Foe at the UN? 3/28/2012. Accessed 8/25/2012, available at: http://www.washingtonpost.com/blogs/fact-checker/post/mitt-romney-and-russia-geopolitical-foe-at-the-un/2012/03/27/gIQATzCneS_blog.htm
18 Martin, Adam. 2012. "Medvedev Mocks Romney in Obama's Place." *The Atlantic Wire* 3/27/2012. Accessed 3/27/2012, available at: http://www.theatlanticwire.com/global/2012/03/medvedev-mocks-romney-obamas-place/50395/
19 Kessler 2012.
20 Popescu, Nicu. 2006. "Outsourcing de-Facto Statehood: Russia and the Secessionist Entities in Georgia and Moldova." *CEPS Policy Briefs* 1:12, 1–8 on www.ceeol.com.
21 Romney, Mitt. 2012. "Bowing to the Kremlin." 3/27/2012. Accessed 8/29/2012, available at: http://www.foreignpolicy.com/articles/2012/03/27/bowing_to_the_kremlin
22 Midlarsky, Manus I. 1975. *On War: Political Violence in the International System* (New York: Free Press).
23 Roberts, Cynthia and Thomas Sherlock. 1999. "Bringing the Russian State Back in: Explanations of the Derailed Transition to Market Democracy." *Comparative Politics* 31 (4): 477–498.
24 Bahry, Donna and Christine Lipsmeyer. 2001. "Economic Adversity and Political Mobilization in Russia" *Electoral Studies* 20 (September 2001): 371–398.
25 Roberts and Sherlock 1999.
26 Goldman, Marshall I. 2008. *Petrostate: Putin, Power, and the New Russia.* (Oxford: Oxford University Press).
27 Goldman 2008.
28 There is no definitive proof that Putin was behind these attacks; however, circumstantial evidence points in his direction.
29 Goldman 2008.
30 Goldman 2008.

31 Goldman 2008.
32 Other oil, gas, and pipeline companies exist alongside the three state-owned giants; therefore, complete state monopolization does not exist in Russia.
33 Aleksashenko, Sergey. 2012. "Russia's Economic Agenda to 2020." *International Affairs* 88 (1): 31–48.
34 Aleksashenko 2012.
35 Goldman 2008, Aleksashenko 2012.
36 Valeriano, Brandon and Ryan C. Maness. 2015. *Cyber War versus Cyber Realities: Cyber Conflict in the International System* (New York: Oxford University Press).
37 Valeriano, Brandon and Ryan C. Maness. 2014. "The Dynamics of Cyber Conflict between Rival Antagonists, 2001–2011." *Journal of Peace Research*, 51 (3): 347–360.
38 Goldman 2008.
39 Russia has been fairly supportive of US air and supply bases, first in Uzbekistan and now Kyrgyzstan. Also when Pakistan closed off the Karachi port, Russia allowed US forces to use its rail lines and airspace for resupply. Russia seems willing to buck pass to the US when it comes to fighting Islamic extremism in Central Asia.
40 We define elites for the purposes of this volume as those Russians in positions of power; either in government, business, media, or the military. In the 21st century, being close to Vladimir Putin is important.
41 Machiavelli, Niccolo. 1513. *The Prince*, George Bull, ed. (New York: Penguin Books, 2003). Morgenthau, Hans J. 1948. *Politics Among Nations* (New York: Knopf).
42 Tsyganov, A.P. 2012. "The Heartland No More: Russia's Weakness and Eurasia's Meltdown." *Journal of Eurasian Studies* 3: 1–9.
43 Nygren 2008.
  Buzan, Barry. 2003. "Security Architecture in Asia, the Interplay of Regional and Global Levels." *The Pacific Review* 16 (2): 143–173.
44 Gabel, Matthew and Kenneth Scheve. 2007. "Estimating the Effect of Elite Communications on Public Opinion Using Instrumental Variables. Accessed 9/16/2012, available at: http://www.yale.edu/leitner/resources/docs/elite-01-07.pdf
45 Diehl, Paul F. 1992. "What Are They Fighting For? The Importance of Issues in International Conflict Research." *Journal of Peace Research* 29 (3): 333–344.
  Hensel, Paul R. 2001. "Contentious Issues and World Politics: The Management of Territorial Claims in the Americas, 1816–1992. *International Studies Quarterly* 45: 81–109.
46 Feklyunina, Valentina. 2008. "Battle for Perceptions: Projecting Russia in the West." *Europe Asia Studies* 60 (4): 605–629.
  Feklyunina, Valentina. 2012. "Russia's International Images and its Energy Policy. An Unreliable Supplier?" *Europe Asia Studies* 64 (3): 449–469.
47 Feklyunina 2008, 2012.
48 Becker, Michael A. 2010. "Russia and the Arctic: Opportunities for Engagement within the Existing Legal Framework." *American University International Law Review* 25 (2): 225–250.

Chapter 2

1 Rose, Gideon. 1998. "Neoclassical Realism and Theories of Foreign Policy." *World Politics* 51 (1):144–17 http://www.ourdocuments.gov/doc.php?flash=true&doc=232.
2 Monroe Doctrine. Accessed 9/17/2012, available at: http://www.ourdocuments.gov/doc.php?flash=true&doc=23
3 Vasquez, John A. 1986. *Evaluating U.S. Foreign Policy* (New York: Praeger): 3.
4 Aslund, Anders. 2012. "Why Moldova Has Turned Its Back on Russia." *The Moscow Times* 7/26/2012. Accessed 7/27/2012, available at: http://www.themoscowtimes.com/opinion/article/why-moldova-has-turned-its-back-on-russia/462640.html
5 Aslund 2012.
6 The state must also have active and past rivalries, the ability to project power, and the capacity to negotiate through international institutions.
7 Lucas, Edward. 2008. *The New Cold War: Putin's Russia and the Threat to the West* (New York: Palgrave Macmillan).
8 Kubicek, Paul J. 2004. "Russian Energy Policy in the Caspian Basin." *World Affairs* 166 (4): Spring 2004: 207–219.
9 Mankoff, Jeffrey. 2009. *Russian Foreign Policy: The Return of Great Power Politics* (New York: Rowman and Littlefield).
10 Kubicek, Paul J. 2001. "Regionalism in Post-Soviet Ukraine," in T. Clark and D. Kempton, eds. *Unity or Separation: Center-Periphery Relations in the Former Soviet Union* (New York: Praeger): 227–249.
11 Sherr, James. 2013. *Hard Diplomacy and Soft Coercion: Russia's Influence Abroad* (London: Chatham House).
12 Averre, Derek L. 2009. "Competing Rationalities: Russia, the EU and the 'Shared Neighbourhood.'" *Europe-Asia Studies,* 61 (10): 1689–1713.
13 Allison, Roy. 2013. *Russia, the West, and Military Intervention* (Oxford: Oxford University Press).
14 Ambrosio, Thomas. 2009. *Authoritarian Backlash: Russian Resistance to Democratization in the Former Soviet Union* (London: Ashgate).
15 Kubicek, Paul J. 1999. "Russian Foreign Policy and the West." *Political Science Quarterly* 114 (4): 547–568.
16 Ziegler, Charles. 2008. "Russian and the CIS: Axis of Authoritarianism." *Asian Survey* 49 (1): 135–145.

Ziegler, Charles. 2013. "Energy Pipeline Networks and Trust: The European Union and Russia in Comparative Perspective." *International Relations* 27 (1): 3–29.

Berryman, John. 2010. "Russia, NATO Enlargement, and 'Regions of Privileged Interests.'" In *Russian Foreign Policy in the 21st Century,* Roger E. Kanet, ed. (New York: Palgrave Macmillan): 228–245.
17 Ziegler 2013.
18 Berryman 2010.
19 Tsygankov, Andrei P. 2010. *Russian Foreign Policy: Change and Continuity of National Identity* (New York: Rowman and Littlefield, 2nd Edition).
20 Tsygankov, Andrei P. 2006. "If Not by Tanks, then by Banks? The Role of Soft Power in Putin's Foreign Policy." *Europe-Asia Studies* 58 (7): 1079–1099.

21 Feklyunina, Valentina. 2008. "Battle for Perceptions: Projecting Russia in the West." *Europe Asia Studies* 60 (4): 605–629.
   Feklyunina, Valentina. 2012. "Russia's International Images and its Energy Policy. An Unreliable Supplier?" *Europe Asia Studies* 64 (3): 449–469.
22 Tsygankov 2006.
23 Hopf, Ted. 2005. "Identity, Legitimacy, and the Use of Military Force: Russia's Great Power Identities and Military Intervention in Abkhazia." *Review of International Studies* 31: 225–243.
24 Neumann, Iver B. 2008. "Russia as a Great Power, 1815–2007." *Journal of International Relations and Development* 11: 128–151.
25 Nygren, Bertil. 2008. *The Rebuilding of Greater Russia: Putin's Foreign Policy towards the CIS Countries* (New York: Routledge).
26 Mourtizen, Hans and Anders Wivel. 2012. *Explaining Foreign Policy: International Diplomacy and the Russian-Georgian War* (Boulder, CO: Lynne Rienner).
27 Leng, Russell. 1993. *Interstate Crisis Behavior, 1816–1980: Realism Versus Reciprocity* (Cambridge: Cambridge University Press).
28 Leng 1993: 2.
29 Snyder, Richard Carlton. 1954. *Decision-Making as an Approach to the Study of International Politics* (Princeton: Organizational Behavior Section, Princeton University).
30 Vasquez 1986.
31 Morgenthau, Hans J. 1948. *Politics Among Nations* (New York: Knopf).
32 Vasquez, John A. 1993. *The War Puzzle* (New York: Cambridge University Press): 168.
33 Walt, Stephen. 1987. *The Origins of Alliances* (Ithaca, New York: Cornell University Press).
34 Leng, Russell. 1983. "When Will They Ever Learn? Coercive Bargaining in Recurrent Crises." *Journal of Conflict Resolution* 27(3): 379–419.
   Leng, Russell. 1988. "Crisis Learning Games." *American Political Science Review* 82(1): 179–194.
   Leng 1993.
35 Valeriano, Brandon. 2012. *Becoming Rivals: The Process of Interstate Rivalry Development* (London: Routledge).
36 http://carnegie.ru/publications/?fa=22683
37 Johnston, Alastair Iain. 1995. *Cultural Realism: Strategic Culture and Grand Strategy in Chinese History* (Princeton, N.J.: Princeton University Press).
38 Axelrod, Robert. 1984. *The Evolution of Cooperation* (New York: Basic Books).
39 Valeriano, Brandon and Matthew Powers. 2010. "United States – Mexico: Convergence of Public Policy Views in the Post-9/11 World." *Policy Studies Journal*, 38 (4): 745–775.
40 Valeriano, Brandon and Ryan C. Maness. 2014. "The Dynamics of Cyber Conflict between Rival Antagonists, 2001–2011." *Journal of Peace Research*, 51 (3): 347–360.
41 Vasquez 1993.
42 Valeriano, Brandon and Victor Marin. 2010. "Causal Pathways to Interstate War: A Qualitative Comparative Analysis of the Steps to War Theory." *Josef Korbel Journal of Advanced International Studies* Summer (2):1–26.
   Vasquez 1993.

43 George, Alexander L. 1991. *Forceful Persuasion: Coercive Diplomacy as an Alternative to War* (Washington, D.C.: United States Institute of Peace Press).
44 George 1991.
45 Morgenthau 1948.
46 Leng 1993.
47 Allison, Graham and Philip Zelikow. 1999. *The Essence of Decision: Explaining the Cuban Missile Crisis* (New York: Longman).
48 Valeriano and Marin 2010.
49 Rose 1989: 145.
50 Waltz, Kenneth N. 1979. *Theory of International Politics* (New York: Columbia University Press).
51 Mearsheimer, John J. 2001. *Tragedy of Great Power Politics* (New York: Norton).
52 Valeriano, Brandon. 2009. "The Tragedy of Offensive Realism: Testing Aggressive Power Politics Models." *International Interactions* 35 (2):179–206.
53 Snyder, Jack L. 1991. *Myths of Empire: Domestic Politics and International Ambition* (Ithaca, N.Y.: Cornell University Press).
  Van Evera, Stephen. 1999. *Causes of War: Power and the Roots of Conflict* (Ithaca: Cornell University Press).
54 Morgenthau 1948.
  Machiavelli, Niccolo. 1513. *The Prince*, George Bull, ed. (New York: Penguin Books, 2003).
  Carr, E.H. 1939. *Twenty Years Crisis 1919–1939: An Introduction to the Study of International Relations* (New York: Palgrave Macmillan, 2001).
55 Rose 1989: 146.
56 Rose 1989: 167.
57 Vasquez 1993.
58 Morgenthau 1948.
59 Goldstein, Joshua S. 2011. *Winning the War on War: The Decline of Armed Conflict Worldwide* (New York: Dutton).
  Mueller, John E. 1989. *Retreat from Doomsday: The Obsolescence of Major War* (New York: Basic Books).
  Pinker, Steven. 2011. *The Better Angels of Our Nature: Why Violence Has Declined* (New York: Viking).
60 Rose 1989.
  Waltz 1979.
  Lobell, Steven E., Norrin M. Ripsman, and Jeffrey W. Taliaferro, eds. 2009. *Neoclassical Realism, the State, and Foreign Policy* (Cambridge: Cambridge University Press).
61 Onuf, Nicholas. 1998. Constructivism: A User's Manual. *International Relations in a Constructed World*, 59.
  Berger, P. L. and T. Luckmann. 1991. *The Social Construction of Reality: A Treatise in the Sociology of Knowledge* Vol. 10 (London: Penguin UK).
62 Berger, Peter L. and Thomas Luckmann. 1966. *The Social Construction of Reality* (New York: Doubleday).
63 Snyder 1954.
  Rosenau, James N. 1971. *The Scientific Study of Foreign Policy* (New York: Free Press).
64 Levy, Jack. 2008. "Deterrence and Coercive Diplomacy: The Contributions of Alexander George." *Political Psychology* 29 (4): 537–552.

65 Morgenthau 1948.
66 Valeriano 2012.
67 Diehl, Paul and Gary Goertz. 2000. *War and Peace in International Rivalry* (Ann Arbor: University of Michigan Press).
   Thompson, William. 2001. "Identifying Rivals and Rivalries in World Politics." *International Studies Quarterly* 45: 557–586.
68 Valeriano 2012.
   Diehl and Goertz 2000.
   Vasquez, John and Christopher S. Leskiw. 2001. "The Origins and Warproneness of International Rivalries." *Annual Review of Political Science* 4: 295–316.
   Maoz, Zeev and Ben Mor. 1998. "Learning, Preference Change, and the Evolution of Enduring Rivalries" in P. Diehl, ed. *The Dynamics of Enduring Rivalries* (Urbana: University of Illinois Press).
69 Vasquez, John and Brandon Valeriano. 2009. "Territory as a Source of Conflict and a Road to Peace." In *Sage Handbook on Conflict Resolution*, J. Bercovitch, V. Kremenyuk, and I. W. Zartman, eds. (London: Sage Publications).
70 Mansbach, Richard W. and John A. Vasquez. 1981. *In Search of Theory: A New Paradigm for Global Politics* (New York: Columbia University Press).
71 Hensel, Paul R. 2001. "Contentious Issues and World Politics: The Management of Territorial Claims in the Americas, 1816–1992." *International Studies Quarterly* 45 (1): 81–109.
72 Morgenthau 1948.
73 Diehl and Goertz 2000.
74 Thompson 2001.
75 Huth, Paul, D. S. Bennett, and C. Gelpi. 1992. "System Uncertainty, Risk Propensity, and International Conflict Among the Great Powers." *Journal of Conflict Resolution* 36: 478–517.
76 Huth, Bennett, and Gelpi 1992.
77 Thompson 2001.
   Valeriano 2012.
78 Those states that use coercive diplomacy outside of rivalry are well on the way to becoming rivals in the first place (Valeriano 2012). There is also the issue of rivalry linkages where states engaged in rivalry learn the wrong sorts of behaviors and then push other sorts of engagements to take the form of rivalry interactions.
79 Hutchinson, Marc L. and Douglas M. Gibler. 2007. "Political Tolerance and Territorial Threat: A Cross-National Study." *Journal of Politics* 69 (1): 128–142.
80 Gibler, Douglas M., Marc L. Hutchinson, and Steven V. Miller. 2012. "Individual Identity Attachments and International Conflict: The Importance of Territorial Threat." Publication Forthcoming. http://bama.ua.edu/~dmgibler/pubs/Gibler.Hutchison.Miller.2013.pdf.
81 Valeriano, Brandon and Ryan C. Maness. 2012. "Persistent Enemies and Cybersecurity: The Future of Rivalry in an Age of Information Warfare." In *Cyber Challenges and National Security*, D. Reveron, ed. (Washington, D.C.: Georgetown University Press).
82 Valeriano and Powers 2010.

Chapter 3

1. Diehl, P. and G. Goertz. 2000. *War and Peace in International Rivalry* (Ann Arbor: University of Michigan Press). Thompson, W. 2001. "Identifying Rivals and Rivalries in World Politics." *International Studies Quarterly* 45: 557–586.
2. We generally refer to the rivalry between the Soviet Union (later Russia) and the United States as the US-Russian rivalry. During the Soviet period, the Soviet Union is identified if necessary.
3. It is clear, though, that there was a period of de-escalation in the early 1990s that did not endure. Tensions in the Balkans and the subsequent NATO enlargement eastward, however, contributed to the increase in tensions and reawakening of the rivalry. The Iraq War, and more specifically, Russia's perception of the loss of influence in the aftermath of the Color Revolutions around its periphery, completed the shift from a geopolitical rivalry to a regional one.
4. Diehl and Goertz 2000.
5. Thompson 2001.
6. Huth, P., D. Bennett, and C. Gelpi. 1992. "System Uncertainty, Risk Propensity, and International Conflict Among the Great Powers." *Journal of Conflict Resolution* 36: 478.
7. Huth et al. 1992.
8. Using the Thompson coding, territorial rivals would be spatial, and policy rivals would be both ideological and positional.
9. Senese, P. and J. Vasquez. 2008. *The Steps to War: An Empirical Study* (Princeton: Princeton University Press).
10. Goertz, G. and P. Diehl. 1995. "The Initiation and Termination of Enduring Rivalries: The Impact of Political Shocks." *American Journal of Political Science* 39: 30–52.
11. Goertz, G., B. Jones, and P. Diehl. 2005. "Maintenance Processes in International Rivalries." *Journal of Conflict Resolution* 49(5): 742–769.
12. McGinnis, M. and J. Williams. 1989. "Change and Stability in Superpower Rivalry." *American Political Science Review* 83: 1101–1123.
13. Huth et al. 1992.
14. An opposing view by Colaresi (2001) finds that great power rivalries are likely to terminate during periods of deconcentration after capability shifts due to some great powers' inability to compete. Rivals are also likely to initiate during this period, as new states are more able to compete globally.
15. Bennett, D.S. 1998. "Integrating and Testing Models of Rivalry Duration." *American Journal of Political Science* 42(4): 1200–1232.
16. Cioffi-Revilla, C. 1998. "The Political Uncertainty of Interstate Rivalries: A Punctuated Equilibrium Model" in P. Diehl, ed. *The Dynamics of Enduring Rivalries* (Urbana: University of Illinois Press).
17. Eckstein, H. 1975. "Case Study and Theory in Political Science" in *Handbook of Political Science, Volume 7*, F. Greenstein and N. Polsby, eds. (Reading, Massachusetts: Addison-Wesley Publishing).
18. Bennett (1998) codes the rivalry as existing from 1946 to 1991 and is censored due to the end of data. Diehl and Goertz (2000) code the duration from

1946–86, and Thompson (1998, 2001) from 1945–91. Thompson and Dreyer (2012) code the rivalry as current and continuing.
19 Goertz et al. 2005: 744.
20 Valeriano, B. 2003. Steps to Rivalry: Power Politics and Rivalry Formation. *PhD Dissertation*, Vanderbilt University, Chapter 7.
21 Thompson, W. 1995. "Principal Rivalries." *Journal of Conflict Resolution* 39(2): 195–223.
22 Russia did not see the United States as a principal rival from 1991 to 1998 due to its own internal problems, and the United States did not see Russia as a principal rival from 1999 to 2003 due to its ongoing War on Terror and its concerns in Iraq. Pursuit of the "Star Wars" system prior to 1999 and after the Iraq War demonstrates that Russia remains a principal threat. While the stated concern of the missile defense shield is from "rogue threats" such as North Korea and Iraq, these threats do not have adequate ballistic missile technology to truly threaten any NATO actor. There can be no mistake that the intention of the shield is to protect the West from a resurgent Russia. The placement of the shield only proves this point.
23 Ahmed, Saeed, Greg Botelho, and Marie-Louise Gumuchian. 2014. "20 Questions: What's behind Ukraine's Political Crisis?" *CNN*, 2/20/2014, accessed 7/1/2014, available at: http://www.cnn.com/2014/02/18/world/europe/ukraine-protests-explainer/
24 Walker, Marcus, Matthew Dalton, and Carol E. Lee. 2014. "Europe, US Significantly Expand Sanctions against Russia." *The Wall Street Journal* 7/29/2014, accessed 8/1/2014, available at: http://online.wsj.com/articles/europe-u-s-significantly-expand-sanctions-against-russian-economy-1406666111
25 Rogov, S. 1999. "Russia and the United States at the Threshold of the Twenty-First Century." *Russian Social Science Review* 40(3): 12–49.
26 George, A. L. and A. Bennett. 2005. *Case Studies and Theory Development in the Social Sciences* (Cambridge: Mass., MIT Press).
27 Mayers, T. 1991. *Understanding Weapons and Arms Control: A Guide to the Issues* (Washington, D.C.: Brassey's).
28 Milbank, Dana. 2012. "Nuking Dick Lugar." *Chicago Tribune* 5/7/2012. Accessed 19/19/2012, available at: http://articles.chicagotribune.com/2012-05-07/news/ct-oped-0507-lugar-20120507-6_1_nunn-lugar-act-obama-and-lugar-nuclear-weapons
29 Midlarsky, M., J. Vasquez, and P. Gladkov, eds. 1994. *From Rivalry to Cooperation: Russian and American Perspectives on the Post-Cold War Era* (New York: HarperCollins College Publishers).
30 Levy, J. 1994. "Learning and Foreign Policy: Sweeping the Conceptual Minefield." *International Organization* 48(2): 279–312.
31 Adomeit, H. 1998. "Russian National Security Interests" in R. Allison and C. Bluth, eds. *Security Dilemmas in Russia and Eurasia* (London: Royal Institute of International Affairs).
32 Rogov 1991.
33 Kubicek, Paul. 1999. "Russian Foreign Policy and the West." *Political Science Quarterly* 114(4): 547–568.
34 Vasquez, J. 1994. "Building Peace in the Post-Cold War Era" in M. Midlarsky, J. Vasquez, and P. Gladkov, eds. *From Rivalry to Cooperation: Russia and*

American Perspectives on the Post-Cold War Era (New York City: Longman Publishers).
35 Mansbach, Richard W. and John A.Vasquez. 1981. *In Search of Theory* (Princeton: Princeton University Press).
36 Nygren, Bertil. 2008. *The Rebuilding of Greater Russia: Putin's Foreign Policy towards the CIS Countries* (New York: Routledge).
37 Bakiyev was not as overtly pro-Western as Saakishvili and Yushchenko, but he was vehemently against the corrupt and pro-Russian government of Askar Akayev.
38 Way, Lucan. 2008. "The Real Causes of the Color Revolutions." *Journal of Democracy* 19 (3): 55–69.
39 Cornell, Svante and Frederick Starr, eds. 2009. *The Guns of August 2008: Russia's War with Georgia* (New York: ME Sharpe).
40 Cornell and Starr 2009.
41 Cornell and Starr 2009.
42 Nygren 2008.
43 Chamerlain-Creanga, Rebecca and Lyndon K. Allin. 2010. "Acquiring Assets, Debts and Citizens: Russia and the Micro-Foundations of Transnistria's Stalemated Conflict." *Demokratizatsiya* 18 (4).
44 Chamerlain-Creanga and Allin 2010.
45 Kolstø, Pål and Helge Blakkisrud (accepted for publication). "De facto states and democracy. The case of Nagorno-Karabakh." *Communist and Post-Communist Studies*.
46 Nygren 2008.
47 Ahmed et al. 2014.
48 Walker et al. 2014.
49 Walker et al. 2014.
50 Nygren 2008.
51 Tsygankov, Andrei P. 2010. *Russian Foreign Policy: Change and Continuity of National Identity* (New York: Rowman and Littlefield, 2nd Edition).
52 Isachenkov, Vladimir. 2014. "Putin Cornered over Ukraine." *Stars and Stripes*, 7/31/2014, Accessed 8/4/2014, available at: http://www.stripes.com/news/europe/putin-cornered-over-ukraine-1.296158
53 Protsyk, Oleh and Benedict Harzl, eds. 2013. *Managing Ethnic Diversity in Russia* (London: Routledge).
54 Kubicek 1999.
55 Kubicek 1999: 564.
56 Solana, J. 1999. "NATO's success in Kosovo." *Foreign Affairs* 78(6): 114–120.
57 BBC News. "Confrontation over Pristina airport." 3/9/2000. http://news.bbc.co.uk/2/hi/europe/671495.stm.
58 Nezavisimaya Gazeta. 2/2/2000 pg 1
59 Mendeloff, D. 2008. "Pernicious History as a Cause of National Misperceptions: Russia and the 1999 Kosovo War." *Cooperation and Conflict* 43(1): 31–56.
60 Mendeloff 2008.
61 Bancroft-Hinchley, Timothy. 2011. "Libya: Media Manipulation." *Pravda.ru*. 3/9/20011.
62 Bancroft-Hinchley 2011.
63 Balmasov, Sergei. 2011a. "The West distributes monstrous lies to destroy Syria's Assad." *Pravda.ru*. 7/12/2011.

64 Vinocour, John. 2011. "Three Responses, All Bad, to the Syrian Revolt." *NYTimes.com*. 6/20/2011.
65 Galpin, Richard. 2012. "Russian arms shipments bolster Syria's embattled Assad." *BBC News* 1/30/2012. Accessed 9/13/2012, available at: http://www.bbc.co.uk/news/world-middle-east-16797818
66 Vasilyeva, Nataliya. 2012. "Russia Arms Sales to Syria Halted, Says Official." *Huffington Post* 7/9/2012. Accessed 9/13/2012, available at: http://www.huffingtonpost.com/2012/07/09/syria-crisis-russia-arms-sales-halted_n_1658733.html
67 MacFarquhar, Neil and Anthony Shadid. 2012. "Russia and China Block U.N. Action in Syria." *New York Times* 2/4/2012. Accessed 9/13/2012, available at: http://www.nytimes.com/2012/02/05/world/middleeast/syria-homs-death-toll-said-to-rise.html?pagewanted=1&_r=1&sq=UN%20syria%20vote&st=cse&scp=6
68 Russett, B. and Stam, A. 1998. "Courting Disaster: An Expanded NATO vs. Russia and China." *Political Science Quarterly* 113 (3): 361–382.
69 North Atlantic Treaty http://www.nato.int/docu/basictxt/treaty.htm.
70 Galpin 2012.
71 Vasquez, J. 1993. *The War Puzzle* (Cambridge: Cambridge University Press).
72 Semenyako, Evgeny and Petr Barenboim. 2007. *The Moscow-Bruges Concept of a Single Legal and Rule of Law Space for Europe and Russia* (Moscow: Justitceinform).
73 Semenyako and Barenboim 2007.
74 "Russia sells 24 Sukhoi fighter jets to Venezuela." Accessed 9/17/2012, available at: http://english.people.com.cn/200607/28/eng20060728_287499.html.
"Russia sells weapons to Malaysia." http://english.pravda.ru/russia/economics/19-05-2003/2835-weapons-0
Feifer, Gregory. "Russia Finds an Eager Weapons Buyer in Iran." http://www.npr.org/templates/story/story.php?storyId=6906839
75 Trukhachev, Vadim. 2011. "USA to defend Poland against Russia." *Pravda.ru*. 3/9/2011.
76 Trukhachev, Vadim. 2010. "US Defense Machine Bites into Russia's Back." *Pravda.ru*. 2/15/2010.
77 Balmasov, Sergei. 2011b. "World goes back to pre-war condition of 1939." *Pravda.ru*. 4/1/2011.
78 Editorial Board. 2014. "The West Needs a Strategy to Contain the World's Newest Rogue State – Russia." *The Washington Post*, 7/21/2014. Accessed 81/2014, available at: http://www.washingtonpost.com/opinions/the-west-needs-a-strategy-to-contain-the-worlds-newest-rogue-state--russia/2014/07/21/01021db0-10fc-11e4-8936-26932bcfd6ed_story.html
Motyl, Alexander J. 2014. "Putin is Transforming Mother Russia into a Rogue State." *Al Jazeera*, 7/25/2014. Accessed 8/1/2014, available at: http://america.aljazeera.com/opinions/2014/7/flight-mh17-vladimirputinukrainerussianonstateactors.html
79 Holsti, Ole R. 1992. "Public Opinion and Foreign Policy: Challenges to the Almond-Lippmann Consensus Mershon Series: Research Programs and Debates." *International Studies Quarterly* 36 (4): 439–436.
80 Holsti 1992: 461.
81 Holsti 1992: 461.
82 Holsti 1992: 461.

83 Page, Benjamin I. and Robert J. Shapiro. 1992. *The Rational Public: Fifty Years of Trends in Americans' Policy Preferences* (Chicago: University of Chicago Press).
84 Page and Shapiro 1992.
85 Pew Global Attitudes Project http://www.pewglobal.org/question-search/
86 Putnam, Robert D. 1988. "Diplomacy and Domestic Politics: The Logic of Two-Level Games." *International Organization* 42 (3): 427–460.
87 Bueno de Mesquita, Bruce and Randolph M. Siverson. 1995. "War and the Survival of Political Leaders: A Comparative Study of Regime Types and Political Accountability." *American Political Science Review* 89 (4): 841–855.
88 Mor, Ben D. 1997. "Peace Incentives and Public Opinion: The Domestic Context of Conflict Resolution."*Journal of Peace Research* 34 (2): 197–215.
89 Public Opinion Foundation (POF) or *Fond Obshchestvennoe Mnenie*. 2004–2008. "NATO Expansion: population poll," (April 2004) "NATO Expansion: expert poll," (April 2004) "Russia and NATO: monitoring" (June 2006), http://english.fom.ru. "Otnoshenie k vozmozhnomu vstupleniu Gruzii i Ukrainy v NATO" http://bd.fom.ru/report/map/d081521
90 Colaresi, Michael. 2005. *Scare Tactics: The Politics of International Rivalry* (Syracuse: Syracuse University Press).
91 Goertz, Gary, Bradford Jones, and Paul F. Diehl. 2005. "Maintenance Processes in International Rivalries." *Journal of Conflict Resolution* 49 (5): 742–769.
92 Pew Global Attitudes Project http://www.pewglobal.org/question-search/
Gallup Poll Center http://www.gallup.com/home.aspx
BBC Poll Tracker: Interactive Guide to the Opinion Polls http://news.bbc.co.uk/2/hi/uk_news/politics/8280050.stm
World Public Opinion http://www.worldpublicopinion.org/
Russian Public Opinion Research Center http://wciom.com/
Yuri Levada Analytical Center http://www.levada.ru/eng/
93 Shuster, Simon 2011. "How the War on Terrorism Did Russia a Favor." *Time* 9/19/2011. Accessed 8/26/2011, available at: http://www.time.com/time/world/article/0,8599,2093529,00.html
94 Volkov, Vladimir. 2003. "Putin Condemns Iraq War as an 'Error.'" *World Socialist Website* 3/29/2003. Accessed 8/26/2012, available at: http://www.wsws.org/articles/2003/mar2003/russ-m29.shtml
95 Lieven, Anatol. 2006. "Putin vs. Cheney." *New York Times* 5/11/2006. Accessed 8/26/2012, available at: http://www.nytimes.com/2006/05/11/opinion/11iht-edlieven.html
96 Feller, Ben. 2008. "Bush Opposes Independence for Two Regions in Georgia." *USA Today* 8/25/20008. Accessed 8/26/2012, available at: http://www.usatoday.com/news/washington/2008-08-25-2354800962_x.htm
97 Aron, Leon. 2011. "Everything You Think You Know About the Collapse of the Soviet Union Is Wrong." *Foreign Policy* 8/26/2011. Accessed 8/26/2012, available at: http://www.foreignpolicy.com/articles/2011/06/20/everything_you_think_you_know_about_the_collapse_of_the_soviet_union_is_wrong
98 Klein, James P., Gary Goertz, and Paul F, Diehl. 2006. "The New Rivalry Dataset: Procedures and Patterns." *Journal of Peace Research* 43 (3), 331–348.
99 Feller 2008.

100 Maness, Ryan C. and Brandon Valeriano. 2012. "Russia and the Near Abroad: Applying a Risk Barometer for War." *Journal of Slavic Military Studies* 25 (2).
101 Maness and Valeriano 2012.
102 Saunders 2012.
103 Feller 2008.
104 Von Eggert 2012.
105 Wollner, Adam. 2014. "Poll: Half of Americans See Russia as 'Unfriendly' or Worse." *NPR*, 9/18/2013. Accessed 6/1/2014, available at: http://www.npr.org/blogs/itsallpolitics/2013/09/18/223833619/poll-half-of-americans-see-russia-as-unfriendly-or-worse?ft=1&f=1001&utm_campaign=nprnews&utm_source=npr&utm_medium=twitter
106 Goldman 2008.
107 Norris 2005.
108 Ziegler, Charles E. 2009. "Russia and the CIS in 2008." *Asian Survey* 49 (1): 135–145.
109 Ziegler 2009.
110 Stern, Jonathan. 2006. "The Russian-Ukrainian Gas Crisis of 2006." *Oxford Institute for Energy Studies*.
111 Black, Phil and Zarifmo Aslamshoyeva. 2012. "Russian General Raises Idea of Strike against Missile Shield." *CNN* 5/5/2012. Accessed 8/26/2012, available at: http://articles.cnn.com/2012-05-05/world/world_europe_russia-us-missile-defense_1_missile-shield-defense-shield-russian-missiles?_s=PM:EUROPE
112 Black and Aslamshoyeva 2012.
113 Monaghan, Andrew. 2012. "The *Vertikal:* Power and Authority in Russia." *International Affairs* 88 (1): 1–16.
114 Goertz et al. 2006.
115 Chalupa, Irena. 2014. "Opinion Poll Shows Russian Support for Arming Separatists Militants in Ukraine." *Atlantic Council*, 7/1/2014. Accessed 8/1/2014, available at: http://www.atlanticcouncil.org/blogs/new-atlanticist/opinion-poll-shows-russian-support-for-arming-separatist-militants-in-ukraine
116 Feller 2008.
117 Maness and Valeriano 2012.
118 *Levada Analytical Center's Annual Russian Public Opinion Yearbook 2009.* http://en.d7154.agava.net/sites/en.d7154.agava.net/files/Levada2009Eng.pdf. Date: September 2008.
119 *Levada Analytical Center's Annual Russian Public Opinion Yearbook 2009.* http://en.d7154.agava.net/sites/en.d7154.agava.net/files/Levada2009Eng.pdf.
120 Ziegler 2009.
121 Nygren, Bertil. 2008. "Putin's Use of Natural Gas to Reintegrate the CIS Region." *Problems of Post-Communism* 55 (4): 3–15.
122 Norris 2005.
123 Bancroft, Ian. 2009. "Serbia's Anniversary Is a Timely Reminder." *The Guardian* 3/24/2009. Accessed 8/26/2012, available at: http://www.guardian.co.uk/commentisfree/2009/mar/24/serbia-kosovo
124 Norris 2005.
125 Roberts and Sherlock 1999.
126 Norris 2005.

127 Bancroft 2009.
128 Fawkes, Helen. 2009. "Scars of NATO Bombing Still Pain Serbs." *BBC News* 3/24/2009. Accessed 8/26/2012, available at: http://news.bbc.co.uk/2/hi/europe/7960116.stm
129 Norris 2005.
130 Fawkes 2009.
131 Norris 2005.
132 Norris 2005.
133 Bancroft 2009.
134 Fawkes 2009.
135 Norris 2005.
136 Spyer, Jonathan. 2007. "Europe and Iraq: Test Case for the Common Foreign and Security Policy." *Middle East Review of International Affairs* 11 (2): Article 7.
137 Teather, David. 2004. "Bush Jokes about Search for WMD, but It's No Laughing Matter for Critics." *The Guardian* 3/24/2004. Accessed 8/26/2012, available at: http://www.guardian.co.uk/world/2004/mar/26/usa.iraq
138 Spyer 2007.
139 Anderson 2011.
140 Moffett, Sebastian. 2012. "NATO Underplayed Civilian Deaths, Human Rights Group Claims." *Huffington Post* 5/14/2012. Accessed 8/26/2012, available at: http://www.huffingtonpost.com/2012/05/14/nato-civilian-deaths-nato_n_1513667.html
141 UN Security Council Report on Resolution 1973. 3/17/2011. Accessed 8/26/2012, available at: http://www.un.org/News/Press/docs/2011/sc10200.doc.htm
142 Schmitt, Eric. 2012. "NATO Sees Flaws in Air Campaign against Qaddafi." *New York Times* 4/14/2012. Accessed 8/26/2012, available at: http://www.nytimes.com/2012/04/15/world/africa/nato-sees-flaws-in-air-campaign-against-qaddafi.html?pagewanted=all
143 Schmitt 2012.
144 Moffett 2012.
145 Schmitt 2012.
146 SPIRI website. Accessed 8/17/2012, Available at: www.spiri.org.
147 Von Eggert 2012.
148 *Levada Center* http://www.levada.ru/eng/ Date: June 2012.
149 Vasilyeva, Nataliya. 2012. "Russia Arms Sales to Syria Halted, Says Official." *Huffington Post* 7/9/2012. Accessed 8/26/2012, available at: http://www.huffingtonpost.com/2012/07/09/syria-crisis-russia-arms-sales-halted_n_1658733.html
150 Anischchuk, Alexei and Steve Gutterman 2012. "Russian Ready to Support Annan Mission at UN." *Reuters* 3/20/2012. Accessed 8/26/2012.
151 Goldman 2008.
152 Mankoff, J. 2009. *Russian Foreign Policy: The Return of Great Power Politics* (Lanham, M.D.: Rowman & Littlefield Publishers).
153 Quote attributed to Czar Alexander III in *Washington Post National Weekly* edition. December 20–27, 1999.
154 Vasquez 1994.

155 The military's share of the budget, after falling off from 1991 until 1995, increased to 17 percent of the total budget in 1996 and 19 percent in 1997 (see Bluth 1998). The budget has continued to increase in the 21st century as Russia upgrades its armed forces.
156 Schmitt, Eric and Michael S. Schmidt. 2014. "New Law All but Bars Russian GPS Sites in US." *The New York Times*, 12/28/2013. Accessed 8/1/2014, available at: http://www.nytimes.com/2013/12/29/world/europe/new-law-all-but-bars-russian-gps-sites-in-us.html?_r=1&
157 Elder, Miriam. 2012. "Russia Accuses US of Using 'Cold War Tactics' over Magintsky Act." *Guardian* 12/7/2012. Available at: http://www.guardian.co.uk/world/2012/dec/07/russia-us-cold-war-tactics?CMP=twt_fd.

Chapter 4

1 We find the term "cyber attack" to be misconstruing and inappropriate in that it conflates the tactic to sound similar to a conventional military attack. We prefer the terms "cyber incidents," "cyber operations," and "cyber disputes" to the more popular, but hyperbolic, term. We also limit the use of the term "cyberwar" where possible because we agree with Rid (2011) that cyberwar is not about actual war where deaths result. We focus on the use of the term "cyber conflict."
2 Davis, Joshua. 2007. "Hackers Take Down the Most Wired Country in Europe." *Wired* 15 (9) Accessed 8/2/2012, available at: http://www.wired.com/politics/security/magazine/15-09/ff_estonia?currentPage=all
3 Finn, Peter. 2007. "Cyber Assaults on Estonia Typify a New Battle Tactic." *Washington Post* 5/19/2007. Accessed 8/2/2012, available at: http://www.washingtonpost.com/wp-dyn/content/article/2007/05/18/AR2007051802122.html
4 Image available through Google Images, available at: https://www.google.com/search?num=10&hl=en&site=imghp&tbm=isch&source=hp&biw=1600&bih=767&q=saakashvili+hitler&oq=saakashvili+hitler&gs_l=img.12...5686.20724.0.22518.26.22.4.0.0.0.140.1622.20j2.22.0.epsugrpq1..0.0...1.1.ZKCeY3lMS10
5 Markoff, John. 2008. "Before the Gunfire, Cyberattacks." *New York Times* 8/12/2008. Accessed 8/2/2012, available at: http://www.nytimes.com/2008/08/13/technology/13cyber.html.
6 Swaine, Jon. 2008. "Georgia: Russia 'Conducting Cyber War.'" *The Telegraph* 8/11/2008. Accessed 8/2/2008, available at: http://www.telegraph.co.uk/news/worldnews/europe/georgia/2539157/Georgia-Russia- 1 conducting-cyber-war.htm
7 Russia has also been accused of initiating distributed denial of service (DDoS) cyber incidents on Lithuania in 2008 and Kyrgyzstan in 2009; however, it is not clear if these events were led by government-controlled forces, and thus they do not meet our coding criteria (Valeriano, Brandon and Ryan C. Maness. 2013. "The Dynamics of Cyber Conflict between Rival Antagonists, 2001–2011," Unpublished Manuscript). The world media did not report on these supposed attacks; therefore, they are not included in this analysis.

8 Nye, Joesph. 2011. "Nuclear Lessons for Cyber Security?" *Strategic Studies Quarterly*. Winter: 21.
9 Hersh, Seymour. 2010. "The Online Threat: Should We Be Worried About Cyber War?" *New Yorker*. (November 2010).
10 Valeriano, Brandon and Ryan C. Maness. 2014. "The Dynamics of Cyber Conflict between Rival Antagonists, 2001–2011." *Journal of Peace Research*, 51 (3): 347–360.
11 Rid, Thomas. 2011. "Cyberwar Will Not Take Place." *Journal of Strategic Studies*. First Article: 1–28.
12 Gvodsev, Nicholas K. 2012. "The Bear Goes Digital: Russia and Its Cyber Capabilities." In *Cyberspace and National Security*, Derek Reveron, ed. (Washington: Georgetown University Press).
13 Nye, Joesph. 2011b. *The Future of Power* (New York: Public Affairs): 132–139.
14 Libicki, Martin C. 2007. *Conquest in Cyberspace* (Cambridge: Cambridge University Press): 2.
15 Clarke, Richard A. and Robert K. Knake. 2010. *Cyber War: The Next Threat to National Security and What to Do About It* (New York: HarperCollins, Inc.).
16 Valeriano and Maness 2014.
17 Valeriano and Maness 2014.
18 Carr, Jeffrey. 2010. *Inside Cyber Warfare* (Sebastopol, CA: O'Reilly Media, Inc.): 24.
19 Valeriano and Maness 2014.
20 Clarke and Knake 2010.
21 Clarke and Knake 2010, 289.
22 Valeriano and Maness 2014.
23 Valeriano and Maness 2014.
24 Gvodsev 2012.
25 Gvodsev 2012.
26 Gvodsev 2012: 177.
27 Gvodsev 2012.
28 Gvodsev 2012.
29 Teibel, Amy. "Russian Firm: Iran Victim of another Cyber Attack." *Bloomberg Businessweek*. 5/29/2012. Accessed 8/2/2012, available at: http://www.businessweek.com/ap/2012-05/D9V2CLR01.htm
30 Teibel 2012.
31 Shachtman, Noah. "Russia's Top Cyber-Sleuth Foils US Spies, Helps Kremlin Pals." *Wired's Danger Room* (Online 7/23/2012). Accessed 8/2/2012, available at: http://www.wired.com/dangerroom/2012/07/ff_kaspersky/
32 Shachtman 2012.
33 Shachtman 2012.
34 Gvodsev 2012.
35 Gvodsev 2012.
36 Gvodsev 2012.
37 Swaine 2008.
38 Liaropoulos, Andrew. 2012. "Power and Security in Cyberspace: Implications for the Westphalian State System."*Academia.edu*. Accessed 9/12/2012, available at: http://piraeus.academia.edu/AndrewLiaropoulos/Papers/1521159/Power_and_Security_in_Cyberspace_Implications_for_the_Westphalian_State_System

39 Clarke and Knake 2010.
40 Gvodsev 2012.
41 Clarke and Knake 2010.
42 Other countries include France, Ukraine, India, Pakistan, and Japan.
43 Clarke and Knake 2010. "Cyberhub Website" Sponsored by *Booz Allen Hamilton*. Accessed 8/1/2012, available at: http://www.cyberhub.com/
44 Clarke and Knake 2010.
45 Clarke and Knake 2010.
46 Valeriano and Maness 2014.
47 Valeriano and Maness 2014.
48 Coalson, Robert. 2009. "Behind the Estonia Cyberattacks." *Radio Free Europe Radio Liberty* 3/6/2009. Accessed 8/3/2012, available at: http://www.rferl.org/content/Behind_The_Estonia_Cyberattacks/1505613.html
49 Coalson 2009.
50 Davis 2007.
51 Gvodsev 2012.
52 Coalson 2009.
53 "110th Congress First Session: HR Res. 397." http://www.gpo.gov/fdsys/pkg/BILLS- 110hres397ih/pdf/BILLS-110hres397ih.pdf
54 "International Conference on Cyber Conflict Webpage." *NATO Cooperative Cyber Defence Centre for Excellence*. Accessed 8/4/2012, available at: http://www.ccdcoe.org/cycon/
55 "International Conference on Cyber Conflict Webpage." *2012 Law and Policy Track*. Accessed 9/12/2012, available at: http://ccdcoe.org/cycon/272.html
56 Coalson 2009.
57 Nazario, Jose. 2008. "DDoS and Security Reports: The Arbor Networks Security Blog." *Arbor Sert Website*. Accessed 8/5/2012, available at: http://ddos.arbornetworks.com/2008/08/georgia-ddos-attacks-a-quick-summary-of-observations/
58 Swaine 2008.
59 Cornell, Svante and Frederick Starr, eds. 2009. *The Guns of August 2008: Russia's War with Georgia* (New York: ME Sharpe).
60 Swaine 2008.
61 Nazario 2008.
62 A new energy grid infiltration had been discovered at the time of this writing and therefore was included in this analysis.
63 Markoff 2008.
64 Griffin, Jennifer. 2008. "Cyberattack Linked to Company of Former Russian Spies." *Fox News* 12/10/2008. Accessed 8/7/2012, available at: http://www.foxnews.com/politics/2008/12/10/cyber-attack-linked-company-russian-spies/
65 Gorman, Siobhan. 2009. "Electricity Grid in US Penetrated by Spies." *Wall Street Journal* 4/8/2009. Accessed 8/7/2012, available at: http://online.wsj.com/article/SB123914805204099085.html
66 Adelmann, Bob. 2014. "Russian Malware Infecting US Energy Grid." *The New American* 7/2/2014, Accessed 8/2/2014, available at: http://www.thenewamerican.com/tech/energy/item/18615-russian-malware-infecting-us-energy-grid
67 Liaropoulos 2012.

68 Valeriano and Maness 2014.
69 Clayton, Mark. 2014. "Massive cyberattacks slam official sites in Russia, Ukraine." *The Christian Science Monitor*, 3/18/2014, Accessed 8/2/2014, available at: http://www.csmonitor.com/World/Security-Watch/Cyber-Conflict-Monitor/2014/0318/Massive-cyberattacks-slam-official-sites-in-Russia-Ukraine
70 Clayton 2014.
71 Valeriano and Maness 2014.

Chapter 5

1 Hamilton, Lee H. 2014."Achieving Long-term Stability in Ukraine Is Key to Navigating Watershed Moment in East-West Relations." *The Huffington Post* 4/3/2014, available at: http://www.huffingtonpost.com/lee-h-hamilton/ukraine-stability_b_5080475.html

   Hennessey, Kathleen and Christi Parsons. 2014. "Obama: Crimea Not Another Cold War but a 'Contest of Ideas.'" *Los Angeles Times* 3/26/2014, available at: http://www.latimes.com/world/worldnow/la-fg-wn-obama-crimea-cold-war-20140326,0,1349858.story#axzz2y47lmnug

   Levgold, Robert. 2014. "The New Cold War." *The Moscow Times* 4/4/2014, available at: http://www.themoscowtimes.com/opinion/article/the-new-cold-war/497466.html

   Payne, Amy. 2014. "Q & A on Ukraine: After Losing Crimea to Russia, What Next?" *Heritage Foundation* 4/5/2014, available at: http://blog.heritage.org/2014/04/05/qa-ukraine-losing-crimea-russia-next/
2 Kyiv Post. "G7 Leaders: Russia's Aggression against Ukraine 'Unacceptable and Violate International Law' *Kyiv Post* 7/31/2014, Accessed 8/1/2014, available at: http://www.kyivpost.com/content/ukraine/g7-leaders-russias-aggression-against-ukraine-unacceptable-and-violate-international-law-358716.html

   Maness, Ryan C. 2013. "Coercive Energy Policy: Russia and the Near Abroad." PhD Dissertation, University of Illinois at Chicago, Chapter 5, 6.
3 Aslund, Anders. 2012. "Why Moldova Has Turned Its Back on Russia." *The Moscow Times* 7/26/2012. Accessed 7/27/2012, available at: http://www.themoscowtimes.com/opinion/article/why-moldova-has-turned-its-back-on-russia/462640.html
4 Nygren, Bertil. 2008. *The Rebuilding of Greater Russia: Putin's Foreign Policy towards the CIS Countries* (New York: Routledge).
5 Aslund 2012.
6 Aslund 2012.
7 Aslund 2012.
8 Morgenthau, Hans J. 1948. *Politics Among Nations: The Struggle for Power and Peace,* Brief Edition (New York: McGraw-Hill).

   Waltz, Kenneth N. 1979. *Theory of International Politics* (New York: McGraw-Hill).

   Mearsheimer, John J. 2001. *The Tragedy of Great Power Politics* (New York: WW Norton & Company).

9 Nye, Joseph S., Jr. 2004. *Soft Power: The Means to Success in World Politics* (Cambridge: MA: Perseus Book Group).
   Tsygankov, Andrei P. 2006. "If Not by Tanks, then by Banks? The Role of Soft Power in Putin's Foreign Policy." *Europe-Asia Studies* 58 (7): 1079–1099.
10 Baldwin, David A. 1985. *Economic Statecraft* (Princeton: Princeton University Press).
11 Nye 2004.
12 Denisov, Andrei and Alexei Grivach. 2008. "The Gains and Failures of an Energy Superpower." *Russia in Global Affairs* 6 (2): 96–108.
13 Woehrel, Steven. 2009. "Russian Energy Policy toward Neighboring Countries." *CRS Report for Congress.* 5/20/2009.
14 Roberts, Cynthia and Thomas Sherlock. 1999. "Bringing the Russian State Back in: Explanations of the Derailed Transition to Market Democracy." *Comparative Politics* 31 (4): 477–498.
15 Roberts and Sherlock 1999.
16 Roberts and Sherlock 1999.
17 "All members have joined the system as a result of negotiation and therefore membership means a balance of rights and obligations. They enjoy the privileges that other member-countries give to them and the security that the trading rules provide. In return, they had to make commitments to open their markets and to abide by the rules—those commitments were the result of the membership (or "accession") negotiations. Countries negotiating membership are WTO 'observers.'" http://www.wto.org/english/thewto_e/whatis_e/tif_e/org3_e.htm
18 Aleksashenko, Sergey. 2012. "Russia's Economic Agenda to 2020." *International Affairs* 88 (1): 31–48.
19 Aron, Leon. 2009. "The Merger of Power and Property." *Journal of Democracy* 20 (2): 66–68.
20 Goldman, Marshall. 2008. *Petrostate: Putin, Power, and the New Russia* (Oxford: Oxford University Press).
21 Aron 2009: 67.
22 Monaghan, Andrew. 2012. "The *Vertikal:* Power and Authority in Russia." *International Affairs* 88 (1): 1–16.
23 Woehrel 2009.
24 Baldwin 1985: Preface.
25 Greyhill Advisers Trade Statistics. Accessed 5/27/2012, available at: http://greyhill.com/trade-statistics
26 Aleksashenko 2012.
27 Woehrel 2009.
28 Aleksashenko 2012.
29 For a detailed analysis of pipeline and assets-for-debts, see Nygren 2008.
30 Kupchinsky, Roman. 2006. "Gazprom: A Troubled Giant." *Radio Free Europe Radio Liberty.* 1/5/2006. Accessed 8/25/2012, Available at: http://www.rferl.org/content/article/1064448.html
31 Gazprom Official Website. "History of Gazprom." Accessed 8/25/2012, available at: http://www.gazprom.com/about/history/
32 Gazprom 2012.
33 Kupchinksy 2006.

34 Goldman 2008.
35 Goldman 2008.
36 Goldman 2008.
37 Goldman 2008.
38 Kupchinsky 2006.
39 Woehrel 2009.
40 Aleksashenko 2012. This ownership has increased as many Near Abroad countries' debt to Gazprom has increased.
41 Woehrel 2009.
42 Woehrel 2009.
43 Armenia, Belarus, Estonia, Georgia, Latvia, and Lithuania.
44 Goldman 2008.
45 Nygren 2008.
   Gelb, Bernard A. 2007. "Russian Natural Gas: Regional Dependence." *CRS Report for Congress* January 5, 2007.
   Smith, Keith C. 2006. "Security Implications of Russian Energy Policies." *CEPS Policy Briefs* Issue 1-12/2006: 1–5 on www.ceeol.com.
   Ziegler, Charles E. 2009. "Russia and the CIS in 2008." *Asian Survey* 49 (1): 135–145.
46 Denisov and Grivach 2008. Russia's "gas weapon" has begun to lose its potency as some pipelines circumventing Russia have come online. Ukraine, for example, has recently cut its purchases of Russian gas and now buys gas from other countries, mainly Azerbaijan, at lower prices.
47 Woehrel 2009.
48 Woehrel 2009.
49 Stern, Jonathan. 2006. "The Russian-Ukrainian Gas Crisis of 2006." *Oxford Institute for Energy Studies*.
50 Tarr, David and Peter Thomson. 2003. "The Merits of Dual Pricing of Russian Natural Gas." Accessed 4/13/2011, available at: http://siteresources.worldbank.org/INTRANETTRADE/Resources/Tarr&Thomson_DualPricingGasRussia.pdf
51 Tarr and Thomson 2003.
52 Feklyunina, Valentina. 2012. "Russia's International Images and Its Energy Policy. An Unreliable Supplier?" *Europe Asia Studies* 64 (3): 449–469.
53 Mansbach, Richard and John A. Vasquez. 1981. *In Search of Theory* (New York: Columbia University Press).
54 Hubert, Franz and Svetlana Ikonnikova. 2003. "Strategic Investment and Bargaining Power in Supply Chains: A Shapely Value Analysis of the Eurasian Gas Market." First Draft: hubert@wiwi.hu-berlin.de and sveticon@mailru.com.
55 Woehrel 2009.
56 Woehrel 2009.
57 Nygren 2008.
58 Woehrel 2009.
59 Gratz 2010.
60 Yearly data is not available at this time. We could only attain data for the year 2005.
61 Gazprom Official Website. "Gazprom Marketing in the CIS and Baltic States." Accessed 4/13/2011, available at: http://gazprom.com/marketing/cis-baltia/

62. Vedler, Sulev. 2012. "Estonia's Battle for Baltic LNG Terminal." *The Baltic Times Online* May 17, 2012: http://www.baltictimes.com/news/articles/31270/
63. Woehrel 2009.
64. Woehrel 2009.
65. Nygren 2008.
66. Woehrel 2009.
67. Woehrel 2009.
68. Herszenhorn, David M. 2013. "Ukraine in Turmoil after Leaders Reject Major EU Deal." *The New York Times*, 11/26/2013. Accessed 8/1/2014, available at: http://www.nytimes.com/2013/11/27/world/europe/protests-continue-as-ukraine-leader-defends-stance-on-europe.html?pagewanted=all
69. Herszenhorn 2013.
70. Marson, James. 2013. "Russia Bails out Ukraine in Rebuke to US, Europe." *The Wall Street Journal*, 12/17/2013. Accessed 8/1/2014, available at: http://online.wsj.com/news/articles/SB10001424052702304403804579263963348323966
71. Krukowska, Ewa, Elena Mazneva, and Volodymyr Verbyany. 2014. "Ukraine Rejects Gas Offer as Talks End Without Deal." *Bloomberg*, 6/11/2014. Accessed 8/1/2014, available at: http://www.bloomberg.com/news/2014-06-11/russia-ukraine-gas-talks-resume-as-putin-and-merkel-speak.html
72. Kalicki, Jan H. 2004. "Caspian Energy at the Crossroads." *Foreign Affairs* 80 (5): 120–134.
73. Woehrel 2009.
74. Garbe, Folkert, Felix Hett, and Rainer Linder. 2011. "Brothers to Neighbors: Russia-Belarus Relations in Transit." In *Russian Energy Security and Foreign Policy*, Adrian Dellecker and Thomas Gomart, eds. (New York: Routledge): 188–202.
75. Garbe et al. 2011.
76. Garbe et al. 2011.
77. Garbe et al. 2011.
78. Garbe et al. 2011.
79. Smith 2006.
80. Baldwin 1985.
81. Svedberg, Marcus. 2007. "Energy in Eurasia: The Dependency Game." *Transition Studies Review* 14 (1): 195–202.
82. Woehrel 2009.
83. Nygren 2008.
84. This pipeline does not supply Georgia with enough gas to completely circumvent Russia and cut its dependence on Russian gas; therefore, Russia has still been able to coerce Georgia.
85. Lake, David A. 2009. *Hierarchy in International Relations* (Ithaca, NY: Cornell University Press).
86. Kindleberger. 1981. "Dominance and Leadership in the International Economy: Exploitation, Public Goods, and Free Rides." *International Studies Quarterly* 25 (5): 242–254.

    Krasner, Stephen D. 1976. "State Power and the Structure of International Trade." *World Politics* 28 (3): 317–347.

87. Ghosn, Faten, Glenn Palmer, and Stuart Bremer. 2004. "The MID3 Data Set, 1993–2001: Procedures, Rules, and Description." *Conflict Management and Peace Science* 21: 133–154.
88. Stern 2006.
89. Bayer, Reşat. 2006. "Diplomatic Exchange Data set, v2006.1." Available at: http://correlatesofwar.org
90. The DCOW set only provides diplomatic data for every five years. This study replicated Bayer's methods to get scores for every country year, as was needed for the analysis.
91. Bayer 2006.
92. Ghosn et al. 2004.
93. Nygren 2008.
94. Nygren 2008.
95. Nygren 2008.
   Henderson, James, Simon Pirani, and Katja Yafimava. 2013. "CIS Gas Pricing: Towards European Netback?" In *The Pricing of Internationally Traded Gas*, Jonathan Stern, ed. (Oxford: Oxford University Press): 178–223.
96. Woehrel 2009.
97. Ziegler 2009.
98. Feklyunina 2012.
99. Nygren 2008.
100. Tsygankov, Andrei P. 2010. *Russian Foreign Policy: Change and Continuity of National Identity* (New York: Rowman and Littlefield, 2nd Edition).

Chapter 6

1. Nygren, Bertil. 2008. *The Rebuilding of Greater Russia: Putin's Foreign Policy towards the CIS Countries* (New York: Routledge).
2. Kimmage, Daniel. 2005. "Uzbekistan: Bloody Friday in the Ferghana Valley." *Radio Free Europe Radio Liberty* 5/14/2005. Accessed 8/7/2013, available at: http://www.rferl.org/content/article/1058869.html
3. Kimmage 2005.
4. Kimmage 2005.
5. Monaghan, Andrew. 2011. "Uzbekistan: Central Asian Key." In *Russian Energy Security and Foreign Policy*, Adrian Dellecker and Thomas Gomart, eds. (New York: Routledge): 121–131.
6. Nygren 2008.
7. Nygren 2008.
8. Nygren 2008: 199.
9. Nygren 2008.
10. Roberts, John. 2011. "After the War: The Southern Corridor." In *Russian Energy Security and Foreign Policy*, Adrian Dellecker and Thomas Gomart, eds. (New York: Routledge): 170–187.
11. Tsygankov, Andrei P. and Matthew Tarver-Wahlquist. 2009. "Dueling Honors: Power, Identity, and the Russia-Georgia Divide." *Foreign Policy Analysis* 5 (4): 307–326.
12. Tsygankov and Tarver-Wahlquist 2009.

13 Valeriano, Brandon. 2012. *Becoming Rivals: The Process of Interstate Rivalry Development* (London: Routledge).
14 Nygren 2008.
15 Feklyunina, Valentina. 2008. "Battle for Perceptions: Projecting Russia in the West." *Europe Asia Studies* 60 (4): 605–629.
    Feklyunina, Valentina. 2012. "Russia's International Images and Its Energy Policy. An Unreliable Supplier?" *Europe Asia Studies* 64 (3): 449–469.
    Tsygankov, Andrei P. 2006. "If Not by Tanks, then by Banks? The Role of Soft Power in Putin's Foreign Policy." *Europe-Asia Studies* 58 (7): 1079–1099.
    Tsygankov, Andrei P. 2010. *Russian Foreign Policy: Change and Continuity of National Identity 2nd Edition* (New York: Rowman and Littlefield).
16 Dellecker, Adrian and Thomas Gomart, eds. 2011. *Russian Energy Security and Foreign Policy* (New York: Routledge).
17 Hensel, Paul R. 2001. "Contentious Issues and World Politics: The Management of Territorial Claims in the Americans, 1816–1992." *International Studies Quarterly* 45 (1):81–109.
18 Hensel 2001.
    Hensel, Paul R., Sara McLaughlin Mitchell, Thomas E. Sowers II, and Clayton L. Thyne. 2008. "Bones of Contention: Comparing Territorial, Maritime, and River Issues." *Journal of Conflict Resolution* 52 (1): 117–143.
19 Hensel 2001.
20 Hensel 2001: 94.
21 Hensel 2001.
22 Ghosn, Faten and Scott Bennett. 2003. "Codebook for the Dyadic Militarized Interstate Incident Data, Version 3.10." Available online: http://correlatesofwar.org.
23 Hensel 2001: 97.
24 Ray, James Lee. 1993. "Wars Between Democracies: Rare, or Nonexistent." *International Interactions* 18 (3):251–276.
25 Ayres, R. William. 2000. "A World Flying Apart? Violent Nationalist Conflict and the End of the Cold War." *Journal of Peace Research* 37 (1): 107–117.
    Jenne, Erin K. 2004. "A Bargaining Theory of Minority Demands: Explaining the Dog that Did Not Bite in 1990s Yugoslavia." *International Studies Quarterly* 48 (4): 729–754.
    Jenne, Erin K., Stephen M. Saideman, and Will Lowe. 2007. "Separatism as a Bargaining Posture: The Role of Leverage in Minority Radicalization." *Journal of Peace Research* 44 (5): 539–558.
    Saideman, Stephen M. 2002. "Discrimination in International Relations: Analyzing External Support for Ethnic Groups." *Journal of Peace Research* 39 (1): 27–50.
    Saideman, Stephen M. and R. William Ayres. 2007. "Pie Crusts and the Sources of Foreign Policy: The Limited Impact of Accession and the Priority of Domestic Constituencies." *Foreign Policy Analysis* 3: 189–210.
26 Hensel 2001.
27 Silitski, Vitali. 2009. "Tools of Autocracy." *Journal of Democracy* 20 (2): 42–46.
28 Kramer, Mark. 2008. "Russian Policy toward the Commonwealth of Independent States: Recent Trends and Future Prospects." *Problems of Post-Communism* 55 (6): 3–19.

29 Members of NATO cannot also be part of GUAM; therefore, the maximum possible score is 13.
30 Ghosn, Faten, Glenn Palmer, and Stuart Bremer. 2004. "The MID3 Data Set, 1993–2001: Procedures, Rules, and Description." *Conflict Management and Peace Science* 21: 133–154.
   Hensel and Mitchell 2013, Marshall 2013.
31 Nygren 2008: 101.
32 GUAM: Organization for Democracy and Economic Development. Accessed 8/22/2013, available at: http://www.guam-organization.org/en/node
33 European Commission European Neighborhood Policy. Accessed 12/8/2013, available at: http://ec.europa.eu/world/enp/policy_en.htm
34 Kramer 2008.
35 Internal Displacement Monitoring Center Website. "Azerbaijan." Accessed 8/21/2013, available at: http://www.internal-displacement.org/countries/azerbaijan
36 Nygren, Bertil. 2011. "Russian Resource Policies towards the CIS Countries." In *Russian Energy Security and Foreign Policy*, Adrian Dellecker and Thomas Gomart, eds. (New York: Routledge): 223–245.
37 Armenia is also part of the EU's Neighborhood Policy and is considered a democratic state according to Polity IV. However, it is also part of the Russian-led CSTO security organization, which keeps its salience score at a low 2.
   Marshall 2013.
38 Nygren 2008.
39 Embassy of the United States in Baku, Azerbaijan. "Economic Data and Reports." Accessed 12/8/2013, available at: http://azerbaijan.usembassy.gov/economic-data.html
40 Holsti, Ole R. 1992. "Public Opinion and Foreign Policy: Challenges to the Almond-Lippmann Consensus. Mershon Series: Research Programs and Debates." *International Studies Quarterly* 36 (4): 439–436.
41 Levada Analytical Center's Annual Russian Public Opinion Yearbook 2009. Accessed 7/31/2012, available at: http://en.d7154.agava.net/sites/en.d7154.agava.net/files/Levada2009Eng.pdf.
   Levada Analytical Center's Annual Russian Public Opinion Yearbook 2011. Accessed 8/14/2013, available at: http://en.d7154.agava.net/sites/en.d7154.agava.net/files/Levada2011Eng.pdf.
   Pew Global Attitudes Project. Accessed 8/21/2013, available at: http://www.pewglobal.org/question-search/
   Russian Public Opinion Research Center. Accessed 19/082013, available at: http://www.wciom.com/index.php?id=61&uid=154
42 Colaresi, Michael. 2005. *Scare Tactics: The Politics of International Rivalry* (Syracuse: Syracuse University Press).
43 Klein, James P., Gary Goertz, and Paul F. Diehl. 2006. "The New Rivalry Dataset: Procedures and Patterns." *Journal of Peace Research* 43 (3), 331–348.
44 Kramer 2008.
45 Tsygankov, Andrei P. 2007. "Finding a Civilizational Idea: 'West,' 'Eurasia,' and 'Euro-East' in Russia's Foreign Policy." *Geopolitics* 12 (3): 1–18.
46 Tsygankov 2010: 1.
47 Ziegler, Charles. 2009. "Russia and the CIS: Axis of Authoritarianism." *Asian Survey* 49 (1): 135–145.

48 Ziegler 2009.
49 Kramer 2008.
50 Feklyunina 2008: 607–608.
51 Tsygankov and Tarver-Wahlquist 2009.
52 Feklyunina 2008: 613–614.
53 Only Azerbaijan and Georgia have pipeline subsidies included in the pricing.
54 Corso, Molly. "Georgia: No Plans to Import Natural Gas from Russia." *Natural Gas Europe* 5/21/2013. Accessed 5/31/2013, available at: http://www.naturalgaseurope.com/georgia-russian-gas-imports
55 Nygren 2011.
56 Ziegler, Charles. 2008. "Competing for Markets and Influence: Asian National Oil Companies in Eurasia." *Asian Perspective* 32 (1): 129–163.
57 Henderson, James, Simon Pirani, and Katja Yafimava. 2012. "CIS Gas Pricing: Towards European Netback?" In *The Pricing of Internationally Traded Gas*, Jonathan Stern, ed. (Oxford: Oxford University Press): 178–223.
58 Nygren 2008.
59 Ziegler 2008.
60 Ipek, Pinar. 2007. "The Role of Oil and Gas in Kazakhstan's Foreign Policy: Looking East or West?" *Europe-Asia Studies* 59 (7): 1179–1199.
61 Ipek 2007.
62 Ziegler 2008.
63 Kimmage 2005.
64 Nygren 2008.
65 Nygren 2011.
66 Ziegler 2009.
67 Kambayashi, Satoshi. "Rising China, Sinking Russia." *The Economist* 9/14/2013. Accessed 10/3/2013, available at: http://www.economist.com/news/asia/21586304-vast-region-chinas-economic-clout-more-match-russias-rising-china-sinking
68 Ipek 2007.
69 Nygren 2008.
70 Tsygankov, Andrei P. 2009 (December). "What Is China to Us? Westernizers and Sinophiles in Russian Foreign Policy." *Russia/NIS Center* 45.
71 Nygren 2008.
72 Nygren 2008.
73 Nygren 2008.
74 Nygren 2008: 162.
75 Nygren 2008: 162.
76 Kislov, Daniil. "Uzbekistan: Farewell to CSTO." 2012 (March 7). Accessed 8/8/2012, available at: http://enews.fergananews.com/article.php?id=2767
77 Nygren 2008.
78 Nygren 2008.
79 Nygren 2008.
80 Henderson et al. 2013.
81 Henderson et al. 2013.
82 Henderson et al. 2013.
83 Dellecker and Gomart 2011.
84 Henderson et al. 2013.

85 The Trans-Afghan pipeline that is to travel from Turkmenistan through Afghanistan to Pakistan and India has been proposed, but there has been little investment due to the volatility of the territory it will cross, so the project is on hold.
86 Dellecker and Gomart 2011.

Chapter 7

1 Dodds, Klaus. 2010. "Flag Planting and Finger Pointing: The Law of the Sea, the Arctic and the Political Geographies of the Continental Shelf." *Political Geography* 29: 63.
2 International Treaty on the Antarctic stipulates that no state can make territorial claims on the continent, as Antarctica is considered a global "commons" and is open to any country that wishes to explore it. This does not apply to the Arctic for reasons indicated on page 35–36.
3 Dodds 2010.
4 Eckel, Mike. 2007. "Russia Defends North Pole Flag-Planting." *USA Today*. Accessed 7/18/2012, available at: http://www.usatoday.com/tech/science/2007-08-08-russia-arctic-flag_N.htm
5 Hensel, Paul R. 2001. "Contentious Issues and World Politics: The Management of Territorial Claims in the Americas, 1816–1992." *International Studies Quarterly* 45: 81–109.

   Vasquez, John A. 1993. *The War Puzzle* (Cambridge: Cambridge University Press).

   Senese, Paul D. and John A. Vasquez. 2008. *The Steps to War Theory: An Empirical Analysis* (Princeton: Princeton University Press).
6 Vasquez 1993.
7 Borgerson, Scott G. 2008. "Arctic Meltdown." *Foreign Affairs* (March/April): 63–77.

   Cohen, Ariel. 2010. "From Russian Competition to Natural Resources Access: Recasting US Arctic Policy." *Backgrounder* 2421 (June 15): 1–13.

   Huebert, Rob. 2010. "Welcome to a New Era of Arctic Security." *Canadian Defence and Foreign Affairs Institute* (August).
8 Borgerson 2008, Cohen 2010, Huebert 2010.
9 Forbes, Bruce C. and Florian Stammler. 2009. "Arctic Climate Change Discourse: The Contrasting Politics of Research Agenda in the West and Russia." *Polar Research* 29: 28–42.
10 Forbes and Stammler 2009.
11 Crawford, Alec, Arthur Hanson, and David Runnalls. 2008. "Arctic Sovereignty and Security in a Climate Changing World." *International Institute for Sustainable Development*. Accessed 3/21/2012, available at: http://www.iisd.org
12 Blunden, Margaret. 2012. "Geopolitics and the Northern Sea Route." *International Affairs* 88 (1): 115–129.
13 Blunden 2012.
14 Young, Oran R. 2009. "The Arctic in Play: Governance in a Time of Rapid Change." *International Journal of Marine and Coastal Law* 24: 423–442.
15 Borgerson 2008.

16 Borgerson 2008.
17 Borgerson 2008.
18 Alexandrov, Oleg. 2009. "Labyrinths of Arctic Policy." *Russia in Global Affairs* 7 (3): 110–118.
19 Huth, Paul. 1996. *Standing Your Ground: Territorial Disputes and International Conflict* (Ann Arbor: University of Michigan Press).
20 Homer-Dixon, Thomas F. 1999. *Environment, Scarcity, and Violence* (Princeton: Princeton University Press).
21 Borgerson 2008, Cohen 2010, Huebert 2010.
22 Mansbach, Richard W. and John A. Vasquez. 1981. *In Search of Theory* (New York: Columbia University Press).
23 Becker, Michael A. 2010. "Russia and the Arctic: Opportunities for Engagement within the Existing Legal Framework." *American University International Law Review* 25 (2): 225–250.
24 Cover, Matt. 2012. "GOP Senators Sink Law of the Sea Treaty; 'This Threat to Sovereignty.'" *CNSNews.com* (July 6). Accessed 9/12/2012, available at: http://cnsnews.com/news/article/gop-senators-sink-law-sea-treaty-threat-sovereignty
25 Ebinger, Charles A. and Evie Zambetakis. 2009. "The Geopolitics of Arctic Melt." *International Affairs* 85 (6): 1215–1232.
26 Titley, David W. and Courtney St. John. 2010. "Arctic Security and the US Navy's Roadmap for the Arctic." *Naval War College Review* 63 (2): 34–48.
27 United Kingdom, Germany, France, Netherlands, Sweden, and Finland.
28 "Arctic Council Website." Accessed 7/20/2012, available at: http://www.arctic-council.org/index.php/en/
29 Baker, Betsy. 2010. "Law, Science, and the Continental Shelf: The Russian Federation and the Promise of Arctic Cooperation." *American University International Law Review* 25 (2): 251–281.
30 Baker 2010.
31 Baker 2010.
32 Becker 2010.
33 http://www.eeas.europa.eu/north_dim/
34 Vasquez 1993.
35 Carmen, Jessie C. "Economic and Strategic Implications of Ice-Free Arctic Seas." In *Globalization and Maritime Power*. Accessed 6/17/2009, available at: http://www.ndu.edu/inss/books/books_2002/Globalization_and_Martime_Power
36 Borgerson 2008, Cohen 2010, Huebert 2010.
37 Huebert 2010.
38 Huebert 2010: 1.
39 Mansbach and Vasquez 1981.
40 Mansbach and Vasquez 1981.
41 Mansbach and Vasquez 1981: 31.
42 Mansbach and Vasquez 1981.
43 Mansbach and Vasquez 1981: 47.
44 Mansbach and Vasquez 1981
45 Mansbach and Vasquez 1981: 61.
46 Diehl, Paul F. 1992. "What Are They Fighting For? The Importance of Issues in International Conflict Research." *Journal of Peace Research* 29 (3): 333–344.
47 Hensel 2001.

48 Hensel 2001.
   Hensel, Paul R., Sara McLaughlin Mitchell, Thomas E. Sowers II, and Clayton L. Thyne. 2008. "Bones of Contention: Comparing Territorial, Maritime, and River Issues." *Journal of Conflict Resolution* 52 (1): 117–143
49 Hensel 2001: 94.
50 Ghosn, Faten, and Scott Bennett. 2003. Codebook for the Dyadic Militarized Interstate Incident Data, Version 3.10. Online: http://correlatesofwar.org.
51 Ghosn and Bennett 2003.
52 Mitchell, Sara McLaughlin and Brandon C. Prins. 1999. "Beyond Territorial Contiguity: Issues at Stake in Democratic Militarized Interstate Disputes." *International Studies Quarterly* 43: 169–183.
53 Becker 2010.
54 Blunden 2012.
55 Baker 2010.
56 Blunden 2012.
57 Blunden 2012.
58 Gourley, Julia L. 2012. "Keeping Things Cool in the Arctic." *Institute of the North*. Accessed 7/18/2012, available at: http://www.institutenorth.org/programs/arctic-advocacy-infrastructure/top-of-the-world-telegraph/current-issue/keeping-things-cool-in-the-arctic
59 Stokke, Olav Schram. 2006. "A Legal Regime for the Arctic? Interplay with the Law of the Sea Convention." *Marine Policy* 31: 402–408.
60 Huebert 2010.
61 Klein, James P., Gary Goertz, and Paul F. Diehl. 2006. "The New Rivalry Dataset: Procedures and Patterns." *Journal of Peace Research* 43 (3): 331–348.
62 Mitchell and Prins 1999.
63 Mitchell and Prins 1999.
64 Dodds 2010.
65 Borgerson 2008.
66 Mitchell and Prins 1999.
67 Hensel 2001.
68 Starr, Harvey. 2005. "Territory, Proximity, and Spatiality: The Geography of International Conflict." *International Studies Review* 7(3): 387–406.
69 "Google Maps Website." Accessed 7/19/2012, available at: https://maps.google.com/
70 "Google Maps Website." Accessed 7/19/2012, available at: https://maps.google.com/
71 Blunden 2012.
72 Alexandrov 2009.
73 Alexandrov 2009.
74 Becker 2010.
75 "US Census Bureau Website." Accessed 7/19/2010, available at: http://www.census.gov/population/www/popclockus.html
76 Titley and St. John 2010.
77 Titley and St. John 2010.
78 Huebert 2010.
79 Woods 2012.

80 Huebert 2010.
81 Blunden 2012.
82 Baker 2010.
83 Huebert 2010.
84 "Encyclopedia of the Nations Website." Accessed 7/20/2012, available at: http://www.nationsencyclopedia.com/economies/Americas/Canada.html
85 "Google Maps Website." Accessed 7/19/2012, available at: https://maps.google.com/
86 Vasquez, John A. and Richard W. Mansbach. 1983. "The Issue Cycle: Conceptualizing Long-Term Global Political Change." *International Organization* 37 (2): 257–279.
87 Vasquez and Mansbach 1983.
88 Vasquez and Mansbach 1983.
89 Vasquez and Mansbach 1983.
90 Vasquez and Mansbach 1983.
91 Vasquez and Mansbach 1983: 269–270.
92 Hutchinson, Marc L. and Douglas M. Gibler. 2007. "Political Tolerance and Territorial Threat: A Cross-National Study." *Journal of Politics* 69 (1): 128–142.
93 Gibler, Douglas M., Marc L. Hutchinson, and Steven V. Miller. 2012. "Individual Identity Attachments and International Conflict: The Importance of Territorial Threat." Publication Forthcoming. http://bama.ua.edu/~dmgibler/pubs/Gibler.Hutchison.Miller.2013.pdf
94 Goemans, Hein. 2006. "Territory, Territorial Attachment and Conflict." In *Territoriality and Conflict in an Era of Globalization*, Miles Kahler and Barbara Walter, eds. (New York: Cambridge University Press).
95 Hutchinson and Gibler 2007, Gibler et al. 2012.
96 Ekos Research Associates. 2011. "Rethinking the Top of the World: Arctic Security Public Opinion Survey." Submitted to Canada Centre for Global Security Studies Munk School of Global Affairs/Walter & Duncan Gordon Foundation January 2011. http://munkschool.utoronto.ca/files/downloads/FINAL%20Survey%20Report.pdf
97 Dodds 2010.
98 Stokke 2006.
99 Huebert 2010.
100 Carmen 2009, Dodds 2010, Huebert 2010.

Chapter 8

1 Barkin, Samuel. 2009. "Realism, Prediction, and Foreign Policy." *Foreign Policy Analysis* 5 (3): 233–246.
2 Levy, J. 1994. "Learning and Foreign Policy: Sweeping the Conceptual Minefield." *International Organization* 48(2): 279–312.
3 Vasquez, John A. 1986. *Evaluating US Foreign Policy* (New York: Praeger).
4 Galpin, Richard. 2012. "Russian Arms Shipments Bolster Syria's Embattled Assad." *BBC News* (January 30). Accessed 9/13/2012, available at: http://www.bbc.co.uk/news/world-middle-east-16797818

5 Thompson, William. 1995. "Principal Rivalries." *Journal of Conflict Resolution* 39 (2): 195–223.
6 Valeriano, Brandon and Ryan C. Maness. 2014. "The Dynamics of Cyber Conflict between Rival Antagonists, 2001–2011." *Journal of Peace Research,* 51 (3): 347–360.
7 Aleksashenko, Sergey. 2012. "Russia's Economic Agenda to 2020." *International Affairs* 88 (1): 31–48.
8 Azar, Edward E. 1972. "Conflict Escalation and Conflict Reduction in an International Crisis, Suez 1956." *Journal of Conflict Resolution* 16 (2): 183–201.

Index

9/11, 45, 74, 89, 181

A
Abkhazia, 18, 21, 68, 71, 72, 97, 153, 162, 166–9
accommodationist (policy), 37–8, 45–6, 49–50, 54, 57, 81, 196
advanced persistent threats (APTs), 103
Afghanistan, 21, 72, 74, 109–10, 116, 119, 152–3, 172, 174, 181
air force, 19, 174
Al Qaeda, 181
Alaska, 206–7
American
cyber, 27, 100, 106, 108, 116, 118–19
economic policy (energy investment), 69, 155, 163–5, 171, 175–7, 186
foreign policy (diplomacy), 17, 28–9, 35, 71, 79, 81, 86–90, 208, 212, 214, 221–2
military, 27, 75, 86–8, 181, 194, 203, 221–2
public and elite opinion, 8, 77, 80–90
power, 21–2, 39, 88–9, 179, 192
anarchy, 42, 138
Arctic
Circle, 190, 194, 204–9
Council, 9, 192–5, 198, 212, 214–16, 225
Ocean, 27, 34, 187, 189–192, 194, 205, 207, 209, 213, 217, 224–5
policy, 17
Armenia, 18, 40, 69, 74–75, 133, 144, 146–7, 153, 160–4, 168–171, 223
arms (weapons), 18, 28, 32, 50, 65, 73, 75, 91, 94–6, 188, 200, 221
Asia, 26, 119, 125, 180, 206, 208, 221

assets, 101126, 128–9, 227
assets-for-debts (policy), 18, 123–4, 145, 149
authoritarian, 52, 78, 139, 144, 152, 155, 159, 175, 177, 224
authoritarianism, 39, 84, 126
autonomy, 69, 91–2, 122, 209
Azerbaijan, 18, 21, 68, 71–2, 97, 153, 162, 166, 168–9

B
backdoors (cyber), 103
Baku, 146, 163–4
Baku-Tbilisi-Ceyhan (BTC) pipeline, 68–9, 162–3
Baku-Tbilisi-Erzurum (BTE) pipeline, 69, 137–8, 162–4, 170–2
Baku-Mozdok pipeline, 171
Balkans, 72, 92, 132, 234
Baltic, 74, 79, 82, 99, 113, 132–3, 139, 144, 147, 208–9, 225
Barents Sea, 201
Belarus, 18, 25, 40, 44, 66, 74–5, 132–4, 136–7, 139, 141, 143–5, 147, 149, 160, 177, 223
Belgrade, 64, 92
Berezovsky, Boris, 25
Bering Sea, 207
Berlin Wall, 93
Black Sea, 76, 153
Blue Stream pipeline, 171
Bosnia, 66, 90–1, 96, 222
Bosnian Serbs, 8, 90–1
botnets (cyber), 103, 114, 117
Brazil, 28, 49, 57
Britain (see Great Britain)
Bronze Soldier (cyber dispute), 99, 110–11
Bukhara-Tashkent-Bishkek-Almaty pipeline, 183
Bulgaria, 76, 132, 172
Bush, George W., 94, 192, 200, 220

C

Canada, 34, 189, 193–4, 199–200, 202–6, 208–9, 212–17, 225
capitalism, 23
case study, 62, 67, 188, 191
Caspian Coastal pipeline, 184
Caspian, 68, 164, 172–3, 182, 184
Caucasus, 7, 9, 17, 24, 27, 34, 67–9, 91, 133, 138, 140, 142–3, 146, 152–6, 161–172, 174, 176, 179–82, 184–6, 223
Central Asia, 7, 9, 11, 17, 25, 30, 34, 40, 46, 59, 77, 89, 110, 116, 125, 130, 132–3, 140, 142–4, 146, 148, 152, 154–6, 160–1, 168, 172–186, 223–4, 229
Central Asia Center pipeline, 183
Chechen, 24, 153
Chechnya, 19, 32
Cheney, Dick, 81
Chernomydrin, Victor, 128
China, 21–22, 26, 28–29, 34, 38–40, 47, 73, 75, 94, 96, 102, 106–7, 118–19, 125, 144, 155, 173–86, 190, 103, 201, 203, 206, 213, 222
Chinese National Petroleum Corporation (CNPC), 175–6
Commonwealth of Independent States (CIS), 33, 37, 40, 74, 133, 159, 162, 177–8, 217
citizenship, 21, 71, 141
climate change, 22, 27, 189, 194, 211
Clinton, Bill, 92
coercive
  diplomacy, 7, 10–11, 13, 15–20, 22, 24–6, 28–30, 32–46, 48–50, 52, 54, 56–8, 60, 62, 64, 66, 68, 70, 72, 74, 76–8, 80–4, 86, 88–90, 92–4, 96, 98, 100–102, 104, 106, 108–9, 112–14, 116, 118, 120–87, 188–226
  energy policy, 7, 13, 33, 34, 44, 56, 135–86
  power, 7, 29, 32, 49
Cold War, 1–11, 13, 16–18, 21–3, 27–8, 34–5, 38, 43, 47, 54, 57–8, 60, 62, 63, 65–7, 71, 73, 76, 82, 84–5, 89, 96–7, 107, 116–7, 129, 140, 149, 179, 187, 188, 198–9, 207, 219, 222
  post (Cold War era), 16, 54, 60, 65–6, 71, 76, 96–7, 149, 179, 188, 199, 222
collective security, 74, 160
Collective Security Treaty Organization (CSTO), 75, 160, 171, 173, 177, 223, 250
color revolutions, 64, 68–9, 75, 96, 234
communism, 178
communist, 60, 75, 91
computer, 27, 102, 106
Congress (U.S.), 69, 79, 108
constructivism, 12, 48
constructivist, 29, 33, 47, 156, 168
continental shelf, 27, 190–3, 200–201, 208
Correlates of War (COW), 139, 157
corruption, 125
Crimea, 15, 64, 71, 76, 122, 133, 222
Crimean
  peninsula, 15, 70, 133
  referendum, 120
Cuba, 28, 46, 54
Cuban Missile Crisis, 46, 65, 77
customs union, 177
cyber
  conflict, 17, 33, 100–101, 109, 112–13, 116–21
  dispute, 118
  incident, 102, 114, 116
  security, 16, 101–2, 104, 106
  technology, 100, 106
  war, 101, 105, 225
Czech Republic, 73, 75, 88

D

debt(s), 11, 18, 26, 44, 66, 123, 127, 129–31, 134, 135–6
defacements (cyber), 102–3, 110, 114
defense, 47–8, 67, 75–6, 102–3, 106–8, 120, 152–3, 194, 205
democracy, 23, 40, 64, 67, 73, 125–6, 159, 162
democratic, 23, 39, 52, 64, 75, 96, 112, 126, 137, 139, 158–9, 177, 198, 203, 250
democratization, 62, 67, 97, 134, 154, 180, 196

Denmark, 34, 189, 193–4, 199, 201, 203, 206, 209, 212, 214–16, 225
deterrence, 112, 225
Diplomatic Correlates of War (DCOW), 139, 248
distributed denial of service (DDoS), 99, 102–3, 105, 107, 110–11, 114–15, 120, 222, 241
domestic politics, 19, 30–3, 36, 39, 41, 51, 54, 59–60, 62–3, 65–6, 73, 76, 78–86, 89–90, 97, 106, 129–31, 136, 146, 149, 150, 165, 188, 200, 211, 220, 222
dual pricing (natural gas), 130
Duma, 79

E
Eastern Europe, 17, 67, 74–5, 88–9, 97, 132
economic
crisis, 23
sanctions, 16, 64, 70–1, 95, 112
statecraft, 8–9, 113, 126, 131–4, 137–8, 143, 147, 223
Egypt, 94, 206
election(s), 18, 20, 24, 67–8, 84, 130, 135
empire (Russian former), 18, 31–2, 39, 63, 66, 91, 119, 121–2, 124, 129, 131–2, 138, 149, 168, 180–1, 219–20, 222–4
empirical, 12, 41, 98, 127, 138–9, 185
Energy Charter Treaty, 125, 149
energy policy, 7, 13, 25, 33–34, 44, 121, 123–4, 133, 138, 144–5, 148–52, 155–6, 159, 161–2, 164–70, 172, 177, 179, 181, 185–6, 223–5
energy resources, 75, 149, 194
energy superpower, 121, 125
ENI (energy company), 132
espionage, 105, 120
Estonia, 18, 27, 33, 43, 56, 74, 83, 99, 100, 105, 109–14, 116, 119–20, 130–4, 141, 146–7, 159–60, 222
ethnic Russian(s), 15–16, 18, 33, 64, 70, 71, 90, 112, 124, 135, 141, 143, 145–6, 150, 157, 159, 171, 174, 201, 223

Eurasian Economic Community (EEC), 160–1, 173, 177, 223
European Union (EU), 16, 18, 33, 36–41, 44, 64, 70, 72, 74, 83, 111–13, 122, 123, 125, 130–2, 134–5, 137, 140, 147, 149–50, 160–4, 190, 193–4, 203, 213, 223, 227
EU Neighborhood Policy, 140, 160, 162–4, 223
exploration (resource), 32, 34, 125, 163, 173, 182, 190, 205, 208, 213, 217, 224

F
Federal Security Service (FSB), 27, 44, 102–5, 108
Finland, 192
Flame virus (cyber), 27, 102, 104, 108, 113, 119
foreign policy (Russian), 7–13, 16–22, 25, 29–59, 62–5, 67, 69, 71, 74, 77, 79–86, 88, 89, 90, 97–98, 100–101, 104, 109, 112–13, 116–18, 120–5, 127, 129–30, 136, 138, 140, 143, 155–6, 165, 168, 170, 187–9, 195–6, 210, 214, 218–24
France, 28, 93–4, 243, 253

G
Gaddafi, Muammar, 72, 94
gas (see natural gas)
Gazprom, 25, 36, 72, 105, 122–4, 126–37, 139, 142–6, 148–50, 154–6, 164–5, 169–72, 176, 182–4, 218, 223
GDP, 8, 24–5
genocide, 69, 72, 91, 168
geoeconomic, 163, 170
geopolitical, 12, 20–2, 30, 32, 53, 57, 60–1, 69, 88, 129, 155, 163, 170, 176, 181, 187–8, 199, 221, 225
Georgia, 8–9, 18, 21, 30, 33, 39–41, 43, 56, 60, 64, 67–9, 71, 74–5, 79, 81–4, 87–90, 97, 99–100, 105, 109–10, 114–16, 119–20, 131, 133, 135, 137, 139–41, 143–8, 153–4, 160–3, 165–72, 185–6, 222–3, 225

Germany, 63, 65, 74, 86, 93, 106–7, 132, 164, 253
glasnost, 128
global warming, 45, 56, 187, 194, 198, 211
globalization, 22
Gorbachev, Mikhail, 65, 128
Great Britain, 28, 80, 106–7
great power, 10, 30, 38, 40, 53, 61–3, 70, 79, 95, 98, 123, 131, 140, 148, 150, 153–6, 161–2, 167–8, 172, 180–1, 184–6
Greenland, 189, 201, 203, 206, 209
GUAM, 75, 140, 160, 162–3, 223
Gusinsky, Vladimir, 25

H
hacktivist, 101, 105, 110
Hamas, 119
hardliner, 42, 45–6, 49, 54, 93, 169, 222
Hensel, Paul, 157–8, 197
Hezbollah, 119
Hitler, Adolf, 99, 111, 115
Hu Jintao, 173
human rights, 50, 67, 79, 81, 97, 152–3, 162, 177
Hungary, 72, 74, 132, 172
Hussein, Saddam, 93

I
ice melt (Arctic), 192, 205, 225
Iceland, 192, 194
identity, 10, 16, 29–31, 33, 39–41, 62, 70, 98, 123, 131, 140, 151, 153–4, 156, 161–2, 167–8, 172, 180–1, 184–6
industry, 23, 75, 108–9, 127–8, 207
infiltration(s) (cyber), 103–4, 106, 108–10, 117, 120, 243
institutional, 16, 23, 125, 158, 160, 191, 197–8, 204, 216
institutionalism, 138
institutions, 20, 23, 34, 37, 41, 57, 60, 70, 92, 125–6, 138, 158, 160, 174, 177, 188, 191–2, 198, 200, 214, 224, 226
integration, 23, 148, 150, 153, 160, 162–3, 180

intercontinental ballistic missiles (ICBMs), 198
international
 institutions, 34, 188, 191
 law, 20, 57, 72, 76, 122, 193, 224–5
 norms, 40, 42, 192, 195, 224
 organizations, 150
 political economy, 13
 regimes, 191, 196, 226
 relations, 11–12, 20, 36, 124, 219
International Monetary Fund (IMF), 66, 142, 143, 147
Internet, 21, 25, 27, 44, 102–9, 111–4, 117, 119
intrusion(s) (cyber), 103–4, 170
Iran, 21, 23, 57, 75–6, 88–9, 102, 106–7, 118–9, 164, 170–1, 180–3
Iran-Armenian pipeline, 171
Iraq, 21, 23, 68–9, 72, 75, 81, 88, 90, 93, 97, 110, 115–16, 119, 162, 175, 181, 234–5
Israel, 21, 48, 104, 106–8, 118–19, 222
issue-based approach, 11–12, 60, 67, 80, 188, 191, 195
Issue Correlates of War (ICOW), 157
Italy, 94, 132, 172
Ivanov, Sergey, 153

J
Japan, 57, 76, 190, 193, 203, 213

K
Karimov, Islam, 152–3
Kasperky Labs, 103–5
Kazakhstan, 18, 25, 40, 66, 75, 133, 139, 141, 143–4, 147, 154, 159–60, 173–5, 177–8, 181–6, 223
Kazi-Magomed-Astara-Abadan pipeline, 171
KGB, 24–5, 27, 105, 126, 129
Khodorkovsky, Mikhail, 25
Kiev, 18, 70, 134–5, 148
Kyrgyzstan, 18, 40, 68, 75, 116, 142, 160, 173–4, 178, 181, 183, 229, 241

L
Latvia, 18, 74, 83, 132–3, 147, 160, 246
Lavrov, Sergey, 74, 188

Lebed, Aleksandr, 79
Libya, 8, 21, 72, 79, 83, 90, 93–4, 97, 115, 119
life expectancy (in Russia), 19
Lithuania, 18, 74, 83, 132–3, 147, 160, 241, 246
logic bombs (cyber), 103
Lukashenko, Alexander, 136

M
Mackay, Peter, 187
Magnitsky Act, 97
Mansbach, Richard, 196, 210
maritime, 12, 16–17, 25, 27, 35, 43, 98, 188, 201, 213, 219
market(s), 23, 26, 81, 131–4, 182, 194
market prices, 33, 81, 123, 125, 127, 130, 150, 154, 156
McAfee, 104
media, 25, 76–7, 104–6, 109, 114, 120, 122, 129, 165, 187, 217, 229
Medvedev, Dmitri, 124, 126, 128, 140, 144
membership (organizational), 113–14, 125, 134, 140, 149, 154, 158, 160, 162–4, 173, 223, 225, 245
method(s) (cyber), 16, 43, 47, 73, 99, 102–3, 105, 109, 111, 114, 119
Middle East, 21, 76, 119, 137, 175, 180, 191, 207, 221
militarism, 49
military
  agreements, 66, 73, 191, 215
  expenditures, 8, 28, 74–5
  intervention, 8, 73, 80, 87, 89–90, 95, 99–100, 113–16, 137, 172, 175, 225
  power, 18–19, 28, 40, 42, 43, 48, 63, 74, 76, 82, 84–5, 96, 112–13, 118, 126, 134, 149–150, 181, 194, 197–8, 202, 205, 209, 216, 220–2
Moldova, 18, 36–7, 40, 64, 67, 69, 71, 74–5, 113, 122–3, 130, 133–7, 139, 146–8, 150, 160, 162, 223
Monroe Doctrine, 35
Moscow, 18, 24, 26, 36, 44, 60, 64, 67, 71, 76, 81, 88, 97, 109, 112–13, 116–17, 119, 122, 124, 129, 131, 133–7, 139, 141, 143, 146–50, 160, 163–4, 170, 175–9, 181, 194, 205–7
Muslim(s), 18, 71, 91, 180–1

N
Nabucco pipeline, 155, 162, 172
Nagorno-Karabakh, 18, 68–9, 163, 170
national identity, 39, 41
national interest, 10, 29, 39, 87, 187, 189
national security, 21, 35, 67, 103–5, 107, 117
nationalism, 82, 86, 91, 95, 96, 216
nationalist, 31, 68, 71, 79
NATO, 8, 29, 37, 40–1, 66–7, 71–6, 83, 87–94, 97, 112–14, 134, 140, 147, 160, 168, 194, 202–4, 222, 225, 230, 234–5
NATO-Russia Permanent Joint Council, 66
natural gas, 8–9, 18, 25–7, 33, 39, 44, 69, 72, 87, 113, 122–5, 127–49, 154, 156–66, 169–77, 179–85, 190, 192, 195–6, 204, 207–9, 211, 217, 223–4
natural resources, 44, 129, 187, 189, 192, 194–5, 197, 204, 217
navy (Russian), 19, 27–8, 42, 96, 121, 205–6
Nazarbayev, Nursultan, 174
Nazi, 99, 111
Near Abroad, 13, 18, 35, 40, 51, 55, 82, 90, 96, 100, 109, 116, 119, 124–5, 131, 133, 141, 149, 154, 167, 227
Netherlands, 253
network (computer), 101, 103, 108, 110, 114, 116, 119, 126
Nordstream pipeline, 132, 135, 145, 149
North Korea, 21, 75–6, 107, 118–19, 235
North Pole, 187, 192–3
Northern Sea Route, 27, 190, 200–202, 204, 206, 208, 214
Northwest Passage, 190, 200, 202–4, 208–9, 213–14
Norway, 34, 113, 189, 193–4, 199–201, 203, 206, 208–9, 212, 214–16, 225

248 *Index*

nuclear, 21, 46, 65–6, 75, 88–9, 97, 104, 107–8, 113, 116, 119, 122, 171, 198

O
Obama, Barack, 15, 21–2, 75–6, 89, 192, 200, 220
offshore drilling, 190–1, 205, 208–9, 218
oil company(s), 25, 126, 129
oligarchs, 23, 25, 125, 128–9
online, 72, 99, 101–2, 107, 115
Orange Revolution, 68, 87
Orthodox (religion), 71, 91, 95, 163, 180
Ottoman, 63, 69, 168

P
packet sniffers (cyber), 103
Pakistan, 222, 229
Paris Club, 66
permafrost, 190, 194, 205
Persian (empire), 168
pipeline(s)
  natural gas, 9, 25, 33, 36, 39, 44, 68–70, 123–7, 129–38, 140, 143–5, 149–50, 154–9, 163–6, 169–77, 180–5, 207, 223–4
  oil, 69, 126, 129
Poland, 74, 76, 88, 132
population (decline in Russia), 19
Poroshenko, Petro, 70, 135
post-9/11, 45
post-Cold War, 16, 54, 60, 65–6, 71, 76, 96–7, 149, 179, 188, 199, 222
post-Soviet space, 10, 17–18, 21–2, 25–6, 29–32, 34, 37–41, 44, 51, 64, 67–8, 70, 77–8, 87–90, 96–7, 100, 109, 116, 119, 121, 123, 131, 133, 137, 140, 144–6, 149, 151–7, 159, 161, 164–7, 170, 172, 175–80, 185–6, 188, 219, 221–3
power politics, 10–11, 16–17, 19–20, 29, 35–8, 40, 42–3, 45, 49–50, 56, 58, 96, 101, 109, 119, 121–3, 149, 187, 189, 191, 217, 219–26
prices (natural gas), 8–9, 19, 33, 36, 38, 41, 123–5, 127–8, 130–1, 133–4, 136–50, 154, 156, 166, 169, 170, 175, 177, 182, 184, 185, 223–4
privatization, 23, 108, 128, 180
public opinion
  American, 80, 86
  Russian, 32, 77, 80, 83, 85–9, 93–5, 114, 154, 165, 177–9, 181, 223
Putin, Vladimir, 15, 18, 24–6, 28–9, 39–40, 67, 70, 72, 79–81, 84, 90, 94, 102–3, 105, 121–30, 133, 135, 137, 140, 144, 149–50, 152–3, 159, 173, 184, 219, 221, 225

Q
qualitative, 12, 33, 138, 154
quantitative, 12, 41, 123–4, 138, 142

R
realism, 11, 43, 47–8, 220
recession, 77, 175
recovery (economic), 24
reform, 24–25, 73, 137
revenue(s), 127–8, 130–1, 150, 183, 219
Romania, 72, 76, 172
Rose Revolution, 68, 137, 153
Rosneft, 25, 72, 126, 129, 164
Russian
  coercive diplomacy, 25, 70, 76, 93, 98
  elite(s), 29, 71, 83, 89, 133, 180
  energy policy, 161, 162, 164, 172
  public opinion (see public opinion)
Russo
  Chinese, 161
  Georgian, 39, 74, 110, 114, 140, 170
  Ukrainian, 87, 134–5, 140, 148, 182
  Western, 191

S
Saakashvili, Mikhail, 69, 137
Sarajevo, 71
Sechin, Igor, 126, 129
Serbia, 8, 71, 92, 132, 172
Serbian, 71–2, 92–3
Serbs, 8, 71, 90–3, 95
severity (cyber), 102–3, 108–11, 113, 116–17, 120

Shanghai Cooperation Organization (SCO), 68, 96, 161, 173, 174, 176–7, 223
Siberia, 128
South Caspian pipeline, 172
South Caucasus pipeline, 170–1
South Korea, 76, 106–7, 118–19
South Ossetia, 18, 21, 68, 71–2, 87, 89–90, 97, 114–15, 153, 162, 166, 168–9
South Stream pipeline, 132–3, 135, 145, 149, 155, 172
sovereign(ty), 16, 18, 53, 61, 67–69, 71–72, 79, 93, 112, 122, 125, 129, 130, 153, 163, 166, 169, 171, 190, 192, 200–204, 207–9, 212–13
Soviet Union, 7–8, 17–19, 23, 28–9, 32, 39–40, 44, 53, 56, 63, 67, 69, 71, 73, 77, 86, 89, 91, 94, 99, 103, 109, 122–5, 127–9, 131, 133, 135, 137–41, 143, 145, 147, 149, 151–4, 156–7, 161, 163, 167–8, 180–1, 188–9, 198–9, 217, 221, 224–5, 227–8
St. Petersburg, 25, 126, 128, 205, 209
Stalin, Josef, 111, 153, 168
Strategic Arms Reduction Treaty (START), 10, 65–6, 109
steps-to-war theory, 11–12
Stuxnet worm, 27, 99, 102, 104–5, 108, 113, 119
Sweden, 192
Symantec, 104
Syria, 8, 21, 23, 27–9, 54, 64, 72–3, 75, 79, 83, 90, 94–5, 97, 221

T
Tajikistan, 18, 40, 75, 142, 160, 173–4, 178, 181, 183–4
Tajikistan-China pipeline, 184
Tallinn, 99, 112–14, 147, 222
Tartus, 73
Tashkent, 152, 183
Tbilisi, 68–9, 100, 115, 137, 153, 170
terrorism, 24, 81, 104, 137, 210
terrorist(s), 48, 69, 72–3, 101, 107, 153, 174, 180–1
theory, 12, 17, 19–20, 30, 32–4, 37, 41, 47–9, 56–8, 70, 76, 78–9, 98, 109, 116, 122–3, 125, 127, 138, 154–5, 165–6, 169, 179, 185, 219–20
trade, 22, 27, 37, 123, 127, 135, 138, 142, 150, 169–71, 180, 202, 205–6
Trans-Caspian pipeline, 173, 184
Transneft, 25, 126, 129, 164
transparency, 128, 177
Tulip Revolution, 68
Turkey, 46, 68, 76, 88, 132, 137, 153, 164, 171, 180
Turkmenistan, 40, 74, 142–3, 147, 154, 160, 172–3, 175–6, 178, 181–6
Turkmenistan-China pipeline, 173, 182–3
Turkmenistan-Iran pipeline, 183

U
Ukraine, 8, 15–16, 18, 30, 33, 38, 40, 44, 60, 64, 66–8, 70–1, 74–5, 79, 82–3, 86–8, 90, 97, 113, 120–3, 129–37, 139–41, 143–50, 160, 162, 164, 172, 182, 223
Ukrainian, 15–16, 70, 87, 100, 120, 129, 134–5, 140, 145, 148, 150, 182
United Kingdom (UK), 4, 25, 93
United Nations (UN), 15, 21, 29, 34, 71–3, 75, 94–5, 112, 142–3, 147, 150, 153, 176, 192–3
United Nations Convention on the Law of the Sea (UNCLOS), 192–3, 195, 198, 200, 202–3, 213, 217–8, 225
United Nations Security Council, 21, 29, 72–3, 75, 94
United Russia, 126
United States (US), 4, 7–8, 10–11, 15–22, 25–6, 28–30, 32–4, 37–8, 40–1, 43, 46–7, 54, 56, 59, 61, 63–5, 67–71, 73–87, 89–91, 93–7, 100, 102, 104, 106–9, 111–12, 114–20, 127, 131, 137, 144, 146, 148, 152–5, 161–2, 164–8, 170, 172, 175–9, 181, 184–7, 189, 192–4, 198–200, 206–9, 212, 214–18, 222–3, 225–7
universal, 11, 36–7
Ural Mountains, 128

USSR, 9, 19, 39, 59, 64, 66, 86, 91, 124, 130–1, 141, 155, 159–60, 178, 180, 189, 223
Uzbekistan, 18, 40, 75, 140–3, 145, 147, 152–4, 173–6, 178, 181–3, 185–6, 229

V
vandalism (cyber), 102, 110–11, 222
Vasquez, John, 36, 42, 96–7, 196, 210, 221
Venezuela, 75
Vietnam, 77
virus (cyber), 27, 104, 108, 113
Volga River, 128

W
War on Terror, 69, 96, 152, 162, 175, 235
Warsaw Pact, 66
web (World Wide), 27, 102, 106
West(ern) (civilization), 15–16, 18–19, 21, 23, 25, 29, 33, 35, 37–41, 44, 64, 66–8, 70–4, 82–3, 87–95, 99, 104–6, 111–12, 115, 120–5, 127, 130–2, 134–7, 139–40, 142–50, 153, 159–60, 162–4, 166–73, 175–6, 180, 182, 184–5, 187–8, 191,194–5, 217, 219, 223–5
World Trade Organization (WTO), 22, 125, 149
World War II, 27, 63, 99
worm (cyber), 27, 104, 108, 113

Y
Yanukovych, Viktor, 15, 135
Yeltsin, Boris, 24, 79, 92, 125, 128, 180
Yugoslavia, 72, 91–2
Yukos, 25
Yushchenko, Viktor, 130, 135, 148, 236

Z
zero-sum, 46, 53, 61, 84, 167
Zhirinovsky, Vladimir, 79
zombies (cyber), 103

CPSIA information can be obtained at www.ICGtesting.com
Printed in the USA
LVOW09*1828090715

445632LV00008B/43/P